*Theory and Practice
of Solid Mechanics*

Theory and Practice of Solid Mechanics

Thomas H. Dawson
United States Naval Academy
Annapolis, Maryland

PLENUM PRESS · NEW YORK AND LONDON

Library of Congress Cataloging in Publication Data

Dawson, Thomas H
  Theory and practice of solid mechanics.

  Includes bibliographical references and index.
  1. Mechanics, Applied. 2. Solids. I. Title.
TA350.D37   531   76-26010
ISBN 0-306-30931-9

© 1976 Plenum Press, New York
A Division of Plenum Publishing Corporation
227 West 17th Street, New York, N.Y. 10011

Printed in the United States of America

Preface

This book is intended for use by engineers and scientists who have a need for an introduction to advanced topics in solid mechanics. It deals with modern concepts of continuum mechanics as well as with details of the classical theories of elasticity, thermal elasticity, viscous elasticity, and plasticity of solids. The book assumes no prior knowledge of the mechanics of solids and develops the subject entirely from first principles. Rigorous derivations of governing equations are also followed by applications to a number of basic and practical problems.

Cartesian tensors are used throughout the book to express mathematical concepts in a clear and concise fashion. Chapter 1, accordingly, provides a discussion of this topic for those readers not already familiar with it. This material is then followed by detailed discussions in Chapters 2 and 3 of the kinematics of continuum motion and the fundamental principles of mass conservation and momentum balance. Unlike traditional treatments, this material is first developed for the general large-deformation case and only then restricted to small deformations for use in the usual engineering applications. In this way the reader thus gets a fuller picture of the basic governing relations of solid mechanics.

Chapters 4 and 5 treat theory and problems in classical isothermal elasticity. Again, relations are first developed for large deformations and then reduced for the important case of small deformations. Included in the theory in Chapter 4 are discussions of the modern principle of material indifference and the modern concept of material symmetries. Engineering applications for small deformations are discussed at length in Chapter 5.

Chapters 6 and 7 treat the subject of thermal elasticity and provide general formulations of the first and second laws of thermodynamics as applied to continuous media. Chapter 6 discusses the general theory and includes recent applications of the principle of material indifference and the second law of thermodynamics to the governing thermoelastic relations. Chapter 7 discusses engineering-type applications for the case of small deformations.

Chapters 8 and 9 discuss theory and problems in viscous elasticity in the same spirit as the earlier chapters.

Finally, Chapters 10 and 11 discuss theory and problems in plasticity. General equations describing combined elastic and plastic behavior are discussed in Chapter 10, including modern restrictions arising from the principle of material indifference. Applications are discussed in Chapter 11. This chapter also provides a detailed treatment of slip-line theory in connection with engineering applications.

Because of its attention to both theory and problems, this book should prove useful both to practicing professionals and to advanced students in the various engineering and applied science disciplines. My aim has been to present concise yet rigorous and illuminating derivations of the governing equations in each of the established areas of solid mechanics and to follow these derivations with enough applications to provide a working knowledge of the subject.

T. H. Dawson

Annapolis

Contents

PART I GENERAL PRINCIPLES

Chapter 1
Vectors and Cartesian Tensors

1.1.	Scalars and Vectors	3
1.2.	Coordinate Transformations	5
1.3.	Orthogonality Relations	7
1.4.	Addition of Vectors and Multiplication by a Scalar	8
1.5.	Scalar and Vector Products of Two Vectors	8
1.6.	Definition of Cartesian Tensors	10
1.7.	Addition of Cartesian Tensors	11
1.8.	Multiplication of Cartesian Tensors	12
1.9.	Quotient Rule for Second-Order Tensors	12
1.10.	Symmetric and Antisymmetric Tensors	13
1.11.	Antisymmetric Tensor Components	14
1.12.	Eigenvalues and Eigenvectors of Symmetric Tensors	14
1.13.	Principal Axes of a Symmetric Tensor	16
	Selected Reading	19
	Exercises	20

Chapter 2
Kinematics of Continuum Motion

2.1.	Material and Spatial Variables	21
2.2.	Definitions of Displacement, Velocity, and Acceleration	22

2.3.   Deformation Gradients   23

2.4.   Stretch and Angular Distortion of Line Elements   25

2.5.   Condition for Rigid-Body Motion of Material about a Point .   26

2.6.   Decomposition of Deformation Gradients   27

2.7.   General Motion of Material in the Neighborhood of a Point .   28

2.8.   Approximations Valid for Small Deformations   31

2.9.   Motion in the Neighborhood of a Point for Small Defor-
       mations   32

2.10. Geometric Interpretation of Strain and Rotation Components
       of Small Deformation   34

2.11. Examples of Small Deformation   37

2.12. Unabridged Notation   40

2.13. Cylindrical Polar Coordinates   41

       Selected Reading   45

       Exercises   45

Chapter 3
Governing Equations of Motion

3.1.   Conservation of Mass   47

3.2.   Balance of Linear Momentum   48

3.3.   Balance of Angular Momentum   50

3.4.   Evaluation of Time Derivative of Volume Integral   50

3.5.   Green's Theorem   52

3.6.   The Stress Vector   54

3.7.   The Stress Tensor   56

3.8.   Change of Stress Components with Rigid Rotations   57

3.9.   Local Form of Mass Conservation   58

3.10. Local Form of Linear Momentum Balance   59

3.11. Local Form of Angular Momentum Balance   60

3.12. Some Simple Examples of Stress   61

3.13. Stress Boundary Conditions   63

3.14. Approximations Valid for Small Deformations   65

3.15. Unabridged Notation   66

3.16. Cylindrical Polar Coordinates   67

       Selected Reading   68

       Exercises   69

PART II CLASSICAL ELASTICITY

Chapter 4
Theory of Elasticity

4.1. Constitutive Relations for an Elastic Solid   73
4.2. Restrictions Placed on Constitutive Relations by Principle of Material Indifference   74
4.3. Material Symmetry Restrictions on the Constitutive Relations .   77
4.4. Elastic Constitutive Relations Applicable to Small Deformations   80
4.5. Restriction on Elastic Constants Due to Existence of a Strain Energy Function   81
4.6. Restrictions on Elastic Constants Due to Material Symmetries   82
4.7. Constitutive Relations for Isotropic Elastic Materials   84
4.8. Alternate Form of Elastic Constitutive Relations   86
4.9. Governing Equations for Linear Elastic Deformation of an Isotropic Solid   88
   Selected Reading   89
   Exercises   89

Chapter 5
Problems in Elasticity

5.1. Longitudinal and Transverse Elastic Waves   91
5.2. Static Twisting of Rods and Bars   98
5.3. Saint-Venant's Principle 103
5.4. Compatibility Equations 104
5.5. Plane Strain and Plane Stress 107
5.6. Bending of a Thin Beam by Uniform Loading 110
5.7. Equations for Plane Strain and Plane Stress in Polar Coordinates 113
5.8. Thick-Walled Cylinder under Internal Pressure 115
5.9. Circular Hole in a Strained Plate 118
5.10. Strength-of-Material Formulations 120
5.11. Bending and Extension of Beams 122
5.12. Bending and Extension of Thin Rectangular Plates 129
5.13. Axisymmetric Bending and Extension of Thin Cylindrical Shells 135
   Selected Reading 141
   Exercises 141

PART III THERMAL ELASTICITY

Chapter 6
Theory of Thermal Elasticity

6.1.   First Law of Thermodynamics   147
6.2.   Second Law of Thermodynamics   149
6.3.   Definition of a Thermoelastic Solid   150
6.4.   Restrictions Placed on Constitutive Relations by the Second
       Law of Thermodynamics   150
6.5.   Restrictions Placed on Constitutive Relations by Principle of
       Material Indifference   152
6.6.   Restriction to Small Deformations and Small Temperature
       Changes   153
6.7.   Restriction to Isotropic Materials   155
6.8.   Governing Equations for Linear Thermoelastic Deformation of
       an Isotropic Solid   157
       Selected Reading   158
       Exercises   158

Chapter 7
Problems in Thermal Elasticity

7.1.   Thermoelastic Vibrations   161
7.2.   Periodic Temperature Variation on the Boundary of a Ther-
       moelastic Half-Space   164
7.3.   Plane Strain and Plane Stress Thermoelastic Problems   . . .   167
7.4.   Thermal Stresses in a Thin Elastic Strip   171
7.5.   Plane Strain and Plane Stress Equations in Polar Coordinates .   174
7.6.   Hollow Circular Cylinder with Elevated Bore Temperature   .   175
7.7.   Thermal Effects in Beam Deformations   178
       Selected Reading   182
       Exercises   183

PART IV VISCOUS ELASTICITY

Chapter 8
Theory of Viscous Elasticity

8.1.   Definition of a Standard Viscoelastic Solid   187
8.2.   Restrictions Placed by Principle of Material Indifference   . .   188

8.3.   Restriction to Small Deformations 189
8.4.   Restriction to Isotropic Materials 190
8.5.   Reduction of Constitutive Relations for Special Cases . . . 191
8.6.   Governing Equations for Linear Viscoelastic Deformation of
       an Isotropic Solid 193
       Selected Reading 194
       Exercises 194

Chapter 9
Problems in Viscous Elasticity

9.1.   Free Vibration of a Standard Viscoelastic Solid 197
9.2.   Time-Dependent Uniaxial Response of a Standard Viscoelastic
       Solid 199
9.3.   Hollow Circular Cylinder of Kelvin–Voigt Material Subjected
       to Periodic Bore Pressure 202
9.4.   Viscous Effects in Beam Deformations 205
9.5.   Viscoelastic Correspondence Principle 209
9.6.   Laterally Constrained Bar 211
       Selected Reading 212
       Exercises 213

PART V PLASTICITY

Chapter 10
Theory of Plasticity

10.1. Definition of an Elastic-Plastic Solid 217
10.2. Restrictions Placed by Principle of Material Indifference . . 218
10.3. Restriction to Quasilinear Response Independent of Mean
       Stress 220
10.4. Plastic Constitutive Relations Applicable for Negligible Elastic
       Deformations 222
10.5. Governing Equations 223
       Selected Reading 224
       Exercises 224

Chapter 11
Problems in Plasticity

11.1.   Initial Yielding of a Thin-Walled Tube under Combined
        Tension–Torsion Loading   225
11.2.   Initial Yielding of a Hollow Cylinder under Internal Pressure
        Loading   227
11.3.   Twisting of a Circular Rod   228
11.4.   Plastic Extension of a Cylindrical Bar under Simple Tension
        Loading   231
11.5.   Plane Strain Compression   234
11.6.   Plane Strain Deformation of Rigid–Perfectly Plastic Solids   . .   238
11.7.   Reduction of Plane Strain Equations   240
11.8.   Slip-Line Theory   243
11.9.   Numerical Solutions Using Slip-Line Theory   247
11.10. Wedge Penetration in a Rigid-Plastic Material   254
        Selected Reading   257
        Exercises   257

Appendix A
Similitude and Scale Modeling in Solid Mechanics

Similitude and Scale Modeling in Solid Mechanics . . .   259

Appendix B
Introduction to Numerical Methods in Solid Mechanics

Introduction to Numerical Methods in Solid Mechanics   263

Index   275

THEORY AND PRACTICE
OF SOLID MECHANICS

PART I

GENERAL PRINCIPLES

The subject of solid mechanics is concerned with the motion of solid bodies under the action of applied forces or other disturbances. It deals, in particular, with bodies made of materials like steel or glass, which exhibit definite shape changes under suitably applied forces. As such, it differs from elementary rigid-body mechanics by inquiring not only into the overall motion of bodies of finite size, but also into the relative movement, or deformation, of their individual parts.

The general principles employed in the study of solid mechanics are the same as those used in studying the mechanics of any continuous medium. Although these principles have been the subject of continual study and refinement for over 200 years, their basic formulations can be attributed primarily to the work of two men: *Leonhard Euler* and *Augustin Cauchy*. It was *Euler* who, working in Berlin and St. Petersburg during the eighteenth century, first laid down the general principles of linear and angular momentum balance for continuous media upon which rest all of modern continuum mechanics. His hydraulic researches also led to a full understanding of the principle of conservation of mass and the concept of internal fluid pressure. The work of *Euler* was subsequently extended by the French mathematician *Cauchy* during the early part of the nineteenth century. By introducing the notion of stress at a point, he generalized *Euler's* concept of internal fluid pressure and made it applicable to any continuous medium. *Cauchy* also found the general differential equations of motion of a continuum in terms of the stress and his work on elasticity provided a detailed kinematic theory of deformation.

1

Vectors and Cartesian Tensors

The mathematical concepts associated with the study of solid mechanics, like those of elementary mechanics, may be represented using nothing more than ordinary scalar and vector mathematics. The use of such mathematics has, however, generally been found to be somewhat awkward and lengthy and it has become popular in recent years to employ instead the unified mathematics of *tensors*—and, in particular, *Cartesian tensors* — for representing these concepts in a clearer and more concise fashion. In keeping with this modern trend, Cartesian tensors will therefore be employed throughout this book. We accordingly begin our study of solid mechanics with a brief review of certain ideas from vector analysis as expressed in the tensor notation and subsequently use these ideas to introduce the more general concept of Cartesian tensors.

1.1. Scalars and Vectors

The concepts of scalars and vectors are familar from the study of elementary mechanics. A scalar, it will be recalled, is a quantity characterized by magnitude only and a vector is a quantity characterized by both magnitude and direction. By its very nature, a vector quantity is thus representable by a directed line segment, or arrow, in space whose length is proportional to the magnitude of the vector and whose direction coincides with the direction of the vector quantity itself.

To examine a vector quantity in more detail, let us consider the vector

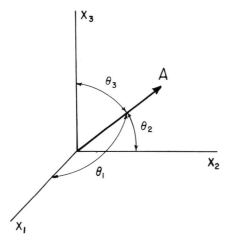

FIGURE 1.1.

A as represented in Figure 1.1. If, as shown, we choose for spatial reference a rectangular Cartesian coordinate system having axes denoted in *index notation* by x_1, x_2, x_3 and having its origin at the base of the arrow representing the vector, it is clear from our definition of a vector that **A** is completely defined by its magnitude A and by its three *direction angles* θ_1, θ_2, and θ_3 specifying its orientation relative to the reference axes x_1, x_2, and x_3.

When the magnitude of the vector and its direction angles are prescribed, it follows also from the geometry of Figure 1.1 that the projections A_1, A_2, A_3 of the vector onto the corresponding coordinate axes x_1, x_2, x_3 are given by the relations

$$A_1 = A \cos \theta_1 = A l_1$$
$$A_2 = A \cos \theta_2 = A l_2 \qquad (1.1)$$
$$A_3 = A \cos \theta_3 = A l_3$$

where l_1, l_2, l_3 denote the *direction cosines* of the corresponding direction angles θ_1, θ_2, θ_3.

Alternatively, if the vector components A_1, A_2, A_3 are specified, the magnitude of the vector and its direction angles can be determined from the following geometrical relations:

$$A = (A_1^2 + A_2^2 + A_3^2)^{1/2}$$
$$\cos \theta_1 = A_1/A$$
$$\cos \theta_2 = A_2/A \qquad (1.2)$$
$$\cos \theta_3 = A_3/A$$

From the above equations, it can be seen that the specification of the components of a vector relative to a reference set of Cartesian coordinate axes is sufficient to define the vector in its entirety. The vector $\mathbf{A}$ introduced above is thus completely represented by the three components A_1, A_2, A_3, which, for sake of brevity, may be written simply as A_i with the understanding, or *range convention*, that any subscript is to take on the values 1, 2, and 3 unless otherwise stated.

1.2. Coordinate Transformations

Consider now a second, primed rectangular Cartesian coordinate system having axes x_1', x_2', x_3' inclined with respect to the original axes x_1, x_2, x_3 as shown in Figure 1.2. If a_{ij} denotes the cosine of the angle between the x_i' and x_j axes shown in Figure 1.2, the components of the vector $\mathbf{A}$ in the primed system are easily seen to be related to those in the unprimed system through the following geometric relations:

$$A_1' = a_{11}A_1 + a_{12}A_2 + a_{13}A_3$$
$$A_2' = a_{21}A_1 + a_{22}A_2 + a_{23}A_3 \qquad (1.3)$$
$$A_3' = a_{31}A_1 + a_{32}A_2 + a_{33}A_3$$

Using the above range convention, these equations may be written more compactly as

$$A_i' = \sum_{j=1}^{3} a_{ij}A_j \qquad (1.4)$$

where i is, of course, understood to take on the values 1, 2, and 3. Also,

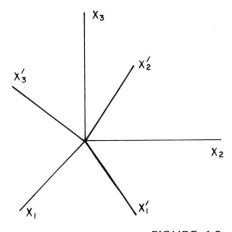

FIGURE 1.2.

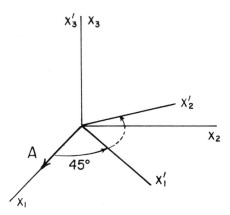

FIGURE 1.3.

since the summation index j in this expression occurs twice, we may achieve a further simplification by adopting the *summation convention* requiring that twice-repeated subscripts in an expression always imply summation over the range 1–3. In this case, equation (1.4) may then be written simply as

$$A'_i = a_{ij}A_j \tag{1.5}$$

It is important to notice that the repeated subscript j in this equation is a so-called dummy index, which can equally well be replaced with another subscript, say k, without changing the meaning of the equation in terms of its expanded form.

The inverse of equation (1.5) may be found immediately from the fact that the equation must remain valid if we interchange primed and unprimed axes. In this case, we have A_i in place of A'_i, A'_j in place of A_j, and $a_{ji} = \cos(x'_j, x_i)$ in place of $a_{ij} = \cos(x'_i, x_j)$ and equation (1.5) becomes

$$A_i = a_{ji}A'_j \tag{1.6}$$

Example. As an example of the use of the above relations, we consider the vector **A** of Figure 1.3, whose components (A_1, A_2, A_3) relative to an unprimed set of axes x_i are $(A, 0, 0)$. We wish to find the components of this vector relative to a set of primed axes x'_i whose orientation differs from the unprimed axes by a rotation of $45°$ about the x_3 axis. The direction cosines a_{ij} are determined from the figure as

$$\begin{bmatrix} a_{11} & a_{12} & a_{13} \\ a_{21} & a_{22} & a_{23} \\ a_{31} & a_{32} & a_{33} \end{bmatrix} = \begin{bmatrix} 1/\sqrt{2} & 1/\sqrt{2} & 0 \\ -1/\sqrt{2} & 1/\sqrt{2} & 0 \\ 0 & 0 & 1 \end{bmatrix}$$

From equation (1.5), the components (A_1', A_2', A_3') of the vector $\mathbf{A}$ relative to the primed set of axes are thus determined immediately as $A/\sqrt{2}$, $-A/\sqrt{2}$, 0).

1.3. Orthogonality Relations

It is worthwhile to consider in greater detail the transformation array $a_{ij} = \cos(x_i', x_j)$ appearing in the above equations. To do this, we introduce the so-called *Kronecker delta* symbol δ_{ij} defined as

$$\delta_{ij} = \begin{cases} 1 & \text{if } i = j \\ 0 & \text{if } i \neq j \end{cases} \tag{1.7}$$

Using this symbol, any set of vector components A_i may then obviously be written as

$$A_i = \delta_{ij} A_j \tag{1.8}$$

From equations (1.6) and (1.5), we have also that

$$A_i = a_{ki} A_k' = a_{ki} a_{kj} A_j \tag{1.9}$$

so that equation (1.8) may be expressed in the form

$$(a_{ki} a_{kj} - \delta_{ij}) A_j = 0 \tag{1.10}$$

Since the vector components A_j may be chosen arbitrarily, it thus follows that

$$a_{ki} a_{kj} = \delta_{ij} \tag{1.11}$$

In a similar way, starting with $A_i' = \delta_{ij}' A_j'$ in place of equation (1.8), where $\delta_{ij}' = \delta_{ij}$, we may also obtain

$$a_{ik} a_{jk} = \delta_{ij} \tag{1.12}$$

Equations (1.11) and (1.12) are referred to as *orthogonality relations*. The first expresses the trigonometric requirements necessary for the x_i axes to be orthogonal when the x_i' axes are, and the second expresses the requirements for the x_i' axes to be orthogonal when the x_i axes are. Obviously, either equation can be derived from the other by simply reversing the definitions of primed and unprimed axes.

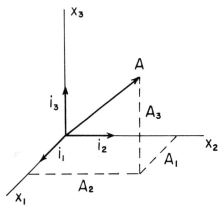

FIGURE 1.4.

1.4. Addition of Vectors and Multiplication by a Scalar

The result of the addition or subtraction of two vectors **A** and **B** is defined to be a third vector **C** having components given by

$$C_i = A_i \pm B_i \tag{1.13}$$

Similarly, the multiplication of a scalar m and a vector **A** is defined to be a vector **C** having components given by

$$C_i = mA_i \tag{1.14}$$

As an example of the use of the above definitions, we may consider unit-magnitude vectors $\mathbf{i}_1$, $\mathbf{i}_2$, $\mathbf{i}_3$ lying along the coordinate axes x_1, x_2, x_3 as shown in Figure 1.4. From the above, it then follows that any vector **A** can be represented mathematically by the equation

$$\mathbf{A} = A_1\mathbf{i}_1 + A_2\mathbf{i}_2 + A_3\mathbf{i}_3 \tag{1.15}$$

where A_1, A_2, A_3 denote the components of the vector **A** along the x_1, x_2, x_3 coordinate directions.

1.5. Scalar and Vector Products of Two Vectors

The scalar product of two vectors **A** and **B** is defined as

$$\mathbf{A} \cdot \mathbf{B} = AB \cos \theta \tag{1.16}$$

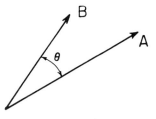

FIGURE 1.5.

where A and B denote the magnitudes of the vectors $\mathbf{A}$ and $\mathbf{B}$ and θ denotes the angle between them as shown in Figure 1.5.

If A_i and B_i denote the components of the vectors $\mathbf{A}$ and $\mathbf{B}$ relative to Cartesian coordinate axes, it follows from equation (1.16) and the vector representations of the form

$$\mathbf{A} = A_1\mathbf{i}_1 + A_2\mathbf{i}_2 + A_3\mathbf{i}_3$$
$$\mathbf{B} = B_1\mathbf{i}_1 + B_2\mathbf{i}_2 + B_3\mathbf{i}_3 \tag{1.17}$$

that the scalar product is expressible as

$$\mathbf{A} \cdot \mathbf{B} = A_1B_1 + A_2B_2 + A_3B_3 \tag{1.18}$$

or, with the summation convention, as

$$\mathbf{A} \cdot \mathbf{B} = A_iB_i \tag{1.19}$$

The vector product of two vectors $\mathbf{A}$ and $\mathbf{B}$ is defined to be a third vector given by the equation

$$\mathbf{A} \times \mathbf{B} = (AB \sin \theta)\mathbf{k} \tag{1.20}$$

where A and B again denote the magnitudes of the respective vectors, θ denotes the smaller angle between them, and $\mathbf{k}$ denotes a unit vector perpendicular to the vectors $\mathbf{A}$ and $\mathbf{B}$, which is postive according to the right-hand rule; that is, positive in the direction in which the thumb points when the right hand curls in the direction formed by rotating $\mathbf{A}$ into $\mathbf{B}$ (see Figure 1.6).

Using the representation of equation (1.17) and the above definition, it is easily found that the vector product can be expressed in terms of the components A_i and B_i by the equation

$$\mathbf{A} \times \mathbf{B} = e_{ijk}A_jB_k\mathbf{i}_i \tag{1.21}$$

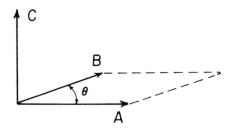

FIGURE 1.6.

where $\mathbf{i}_i$ denotes unit vectors lying along the coordinate axes and where the symbol e_{ijk} is defined as follows:

$e_{ijk} = +1$ for $i = 1$, $j = 2$, $k = 3$ or any even number of permutations of this arrangement (e.g., e_{312})

$e_{ijk} = -1$ for odd permutations of $i = 1$, $j = 2$, $k = 3$ (e.g., e_{132})

$e_{ijk} = 0$ for two or more indices equal (e.g., e_{112})

For example, with $\mathbf{C}$ denoting the vector formed by the operation $\mathbf{A} \times \mathbf{B}$, equation (1.21) yields for the component C_1

$$C_1 = e_{123}A_2B_3 + e_{132}A_3B_2$$

all other contributions to the summation over j and k being zero since they involve cases in which two or more of the indices of e_{ijk} are equal. Also $e_{123} = 1$ and $e_{132} = -1$ since e_{132} can be made equal to e_{123} by *one* interchange of subscripts. Thus C_1 and, similarly, C_2 and C_3 are determined as

$$\begin{aligned} C_1 &= A_2B_3 - A_3B_2 \\ C_2 &= A_3B_1 - A_1B_3 \\ C_3 &= A_1B_2 - A_2B_1 \end{aligned} \qquad (1.22)$$

1.6. Definition of Cartesian Tensors

We now define what is meant by a Cartesian tensor of some specified order. In particular, a Cartesian tensor of order one is defined as a quantity having three components T_i whose transformation between primed and unprimed axes is governed by equations (1.5) and (1.6); that is, by

$$T_i' = a_{ij}T_j \qquad (1.23)$$

and

$$T_i = a_{ji}T'_j \tag{1.24}$$

Thus, a first-order tensor is nothing more than a vector.

Similarly, a Cartesian tensor of order two is defined as a quantity having nine components T_{ij} whose transformation between primed and unprimed coordinate axes is governed by the equations

$$T'_{ij} = a_{ik}a_{jl}T_{kl} \tag{1.25}$$

and

$$T_{ij} = a_{ki}a_{lj}T'_{kl} \tag{1.26}$$

either of which may be obtained from the other by simply reversing definitions of primed and unprimed axes.

Third- and higher- order Cartesian tensors are defined analogously. In a like fashion, it can also be seen that a Cartesian tensor of zeroth order is defined to be any quantity that is unchanged under coordinate transformation; that is, a scalar.

1.7. Addition of Cartesian Tensors

The addition or subtraction of two Cartesian tensors of the same order is defined to be a third Cartesian tensor of the same order having components made up of the sum or difference of the respective components of each. Thus, for example, if A_{ij} and B_{ij} denote components of two second-order tensors, the addition or subtraction of these tensors is defined to be a third tensor of second order having components C_{ij} given by

$$C_{ij} = A_{ij} \pm B_{ij} \tag{1.27}$$

That C_{ij} in the above equation represents, in fact, components of a Cartesian tensor of second order can easily be established by considering the transformation of this equation between primed and unprimed axes. We have, in particular, from equation (1.26) that

$$A_{ij} = a_{mi}a_{nj}A'_{mn}, \qquad B_{ij} = a_{mi}a_{nj}B'_{mn}$$

so that equation (1.27) becomes

$$C_{ij} = a_{mi}a_{nj}(A'_{mn} \pm B'_{mn})$$

or

$$C_{ij} = a_{mi}a_{nj}C'_{mn} \tag{1.28}$$

thus showing that C_{ij} represents components of a second-order Cartesian tensor.

It is worth noticing that when the above definitions are applied to tensors of first order, they reduce to the definitions given earlier for vector addition and subtraction.

1.8. Multiplication of Cartesian Tensors

The multiplication of Cartesian tensors can be classified into two categories, *outer products* and *inner products*. The outer product of two tensors is defined to be a third tensor having components given by the product of the components of the two, with no repeated summation indices. Thus, for example, the outer product of two second-order tensors having components A_{ij} and B_{ij} is defined to be a fourth-order tensor having components C_{ijkl} given by

$$C_{ijkl} = A_{ij}B_{kl} \tag{1.29}$$

That C_{ijkl} represents, in fact, components of a fourth-order Cartesian tensor can easily be established by considering the transformation of equation (1.29) between primed and unprimed axes in a fashion analogous to that used in the previous section for tensor addition and subtraction.

In contrast with the general form of equation (1.29), an inner product of two Cartesian tensors is defined as an outer product followed by a *contraction* of the two; that is, by an equating of any index associated with one tensor to any index associated with the other. For example, an inner product of two tensors having components A_{ij} and B_{kl} can be formed from the outer product of equation (1.29) by equating k and j. We then have the resulting Cartesian tensor of second order whose components C_{il} are given by

$$C_{il} = A_{ij}B_{jl} \tag{1.30}$$

It should be noticed that the scalar and vector products of vectors as defined earlier are nothing more than special cases of the inner product defined here.

1.9. Quotient Rule for Second-Order Tensors

In order to decide whether a set of nine components T_{ij} represents a second-order tensor, we may examine the behavior of these components under coordinate transformation and establish whether or not they satisfy equations (1.25) and (1.26). Alternatively, if we know how the components T_{ij} behave when combined with arbitrary vector components A_i, we can also determine whether or not the components T_{ij} represent a second- order tensor. In particular, suppose we know the following equation to apply:

$$T_{ij}A_i = B_j \qquad (1.31)$$

where B_j denotes components of a vector. Then, the *quotient rule* states that the components T_{ij} are indeed the components of a second-order Cartesian tensor.

To see this, we have only to make use of the above transformation laws. Relative to a set of primed axes, the above equation is

$$T'_{ij}A'_i = B'_j$$

But, with the help of equations (1.23) and (1.31) B'_j can, be written as

$$B'_j = a_{jk}(T_{pk}A_p)$$

Using equation (1.24), this expression may also be written as

$$B'_j = a_{jk}T_{pk}a_{ip}A'_i$$

Hence, equation (1.31) relative to primed axes can be written as

$$(T'_{ij} - a_{ip}a_{jk}T_{pk})A'_i = 0$$

Since the components A'_i are arbitrary, this last equation implies that

$$T'_{ij} = a_{ip}a_{jk}T_{pk} \qquad (1.32)$$

and, hence, from comparison with equation (1.25), that the nine quantities T_{ij} are, indeed, the components of a Cartesian tensor of order two.

1.10. Symmetric and Antisymmetric Tensors

Suppose T_{ij} denotes components of a second-order tensor. We may display these components in the following array:

$$[T_{ij}] = \begin{bmatrix} T_{11} & T_{12} & T_{13} \\ T_{21} & T_{22} & T_{23} \\ T_{31} & T_{32} & T_{33} \end{bmatrix} \qquad (1.33)$$

If the tensor components in the above array are such that $T_{ij} = T_{ji}$, then the tensor is said to be *symmetric*. On the other hand, if the components are such that $T_{ij} = -T_{ji}$, the tensor is said to be *antisymmetric*. Moreover, if a tensor is symmetric (or antisymmetric) when referred to one set of axes x_i, it follows immediately from the tensor transformation law of equation (1.25) that it will also be symmetric (or antisymmetric) when referred to any other set of axes x_i'.

1.11. Antisymmetric Tensor Components

A special characteristic of an antisymmetric tensor is that its operation on a vector is equivalent to an appropriately defined vector-product operation. In particular, if A_i denotes components of a vector and if T_{ij} denotes components of a second-order antisymmetric tensor, then

$$T_{ij}A_j = e_{ijk}W_jA_k \tag{1.34}$$

where W_j denotes vector components defined as follows:

$$W_1 = T_{32}, \qquad W_2 = T_{13}, \qquad W_3 = T_{21} \tag{1.35}$$

That equation (1.34) is valid can easily be established by direct expansion.

1.12. Eigenvalues and Eigenvectors of Symmetric Tensors

Consider the equation

$$T_{ij}A_j = \lambda A_i \tag{1.36}$$

where T_{ij} denotes components of a symmetric tensor, A_i denotes components of a vector, and λ denotes a scalar. Any nonzero vector **A** satisfying this equation is known an an *eigenvector* of the tensor defined by T_{ij} and the associated scalar λ is known as an *eigenvalue*.

Solutions of equation (1.36) can be examined by expanding the equations and rearranging to get

$$
\begin{aligned}
(T_{11} - \lambda)A_1 + T_{12}A_2 + T_{13}A_3 &= 0 \\
T_{21}A_1 + (T_{22} - \lambda)A_2 + T_{23}A_3 &= 0 \\
T_{31}A_1 + T_{32}A_2 + (T_{33} - \lambda)A_3 &= 0
\end{aligned}
\tag{1.37}
$$

The condition for a nontrivial solution of these homogeneous algebraic equations is that the determinant of the coefficients of A_1, A_2, and A_3

vanish; that is, that

$$\begin{vmatrix} T_{11} - \lambda & T_{12} & T_{13} \\ T_{21} & T_{22} - \lambda & T_{23} \\ T_{31} & T_{32} & T_{33} - \lambda \end{vmatrix} = 0 \qquad (1.38)$$

If we regard the components T_{ij} as known, this equation yields the cubic equation

$$\lambda^3 - C\lambda^2 + D\lambda - E = 0 \qquad (1.39)$$

where

$$C = T_{ii}$$
$$D = \tfrac{1}{2} T_{ii} T_{jj} - \tfrac{1}{2} T_{ij} T_{ij}$$
$$E = e_{ijk} T_{i1} T_{j2} T_{k3}$$

When the components T_{ij} are those of a symmetric tensor, it can easily be shown that equation (1.39) will have three real roots. We denote these roots by λ_1, λ_2, and λ_3. Taking first $\lambda = \lambda_1$ in equation (1.37), any two of these three equations can be solved for the ratios A_1/A_3 and A_2/A_3 or, equivalently, l_1/l_3 and l_2/l_3, where l_1, l_2, l_3 denote the direction cosines of the eigenvector associated with the eigenvalue λ_1. Using these two ratios and the geometric relation $l_1^2 + l_2^2 + l_3^2 = 1$, we can then determine the direction of the eigenvector without ambiguity. Taking the length of this eigenvector to be unity, we thus finally have the *unit eigenvector* $\mathbf{e}_1 = l_i \mathbf{i}_i$ associated with the eigenvalue λ_1.

In a similar way, we may also find two additional unit eigenvectors $\mathbf{e}_2$ and $\mathbf{e}_3$ associated with the eigenvalues λ_2 and λ_3.

That the above three unit eigenvectors are mutually perpendicular when λ_1, λ_2, and λ_3 are all distinct can easily be seen from arguments of the following kind: Consider two unit eigenvectors $\mathbf{e}_1$ and $\mathbf{e}_2$. These satisfy equation (1.36) with associated eigenvalues λ_1 and λ_2. Hence,

$$T_{ij} e_{1j} = \lambda_1 e_{1i}, \qquad T_{ij} e_{2j} = \lambda_2 e_{2i} \qquad (1.40)$$

Multiplying the first of these equations by e_{2i} and the second by e_{1i} and subtracting, we have

$$T_{ij} e_{1j} e_{2i} - T_{ij} e_{2j} e_{1i} = e_{1i} e_{2i}(\lambda_1 - \lambda_2)$$

On interchanging the dummy indices i and j in the first term on the left-hand side of this equation and using $T_{ij} = T_{ji}$, we find that

$$e_{1i} e_{2i}(\lambda_1 - \lambda_2) = 0 \qquad (1.41)$$

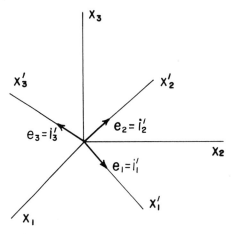

FIGURE 1.7.

Hence, if $\lambda_1 \neq \lambda_2$, then $e_{1i}e_{2i} = 0$. But this is nothing more than the scalar product of the unit eigenvectors $\mathbf{e}_1$ and $\mathbf{e}_2$, so that $\mathbf{e}_1$ and $\mathbf{e}_2$ are therefore perpendicular.

A similar argument shows also that $\mathbf{e}_1$ and $\mathbf{e}_3$ and that $\mathbf{e}_2$ and $\mathbf{e}_3$ are also perpendicular provided $\lambda_1 \neq \lambda_3$ and $\lambda_2 \neq \lambda_3$, respectively.

1.13. Principal Axes of a Symmetric Tensor

Under the above circumstances, we may obviously choose a new set of Cartesian axes x_i' having unit vectors along these axes coincident with the unit eigenvectors; that is, $\mathbf{i}_1' = \mathbf{e}_1$, $\mathbf{i}_2' = \mathbf{e}_2$, and $\mathbf{i}_3' = \mathbf{e}_3$, as shown in Figure 1.7. For this system of axes, we have from equation (1.36) that

$$
\begin{aligned}
T_{ij}'e_{1j}' &= \lambda_1 e_{1i}' \\
T_{ij}'e_{2j}' &= \lambda_2 e_{2i}' \\
T_{ij}'e_{3j}' &= \lambda_3 e_{3i}'
\end{aligned}
\tag{1.42}
$$

Remembering that $e_{1j}' = \delta_{1j}$, $e_{2j}' = \delta_{2j}$, and $e_{3j}' = \delta_{3j}$, we thus find from the first of these equations that

$$
T_{11}' = \lambda_1, \qquad T_{21}' = T_{31}' = 0
\tag{1.43}
$$

and, similarly, from the remaining two that

$$
\begin{aligned}
T_{22}' &= \lambda_2, & T_{12}' &= T_{32}' = 0 \\
T_{33}' &= \lambda_3, & T_{13}' &= T_{23}' = 0
\end{aligned}
\tag{1.44}
$$

In this system of so-called *principal axes* defined by the unit eigenvectors e_1, e_2, e_3, the tensor components are therefore expressible as

$$[T'_{ij}] = \begin{bmatrix} \lambda_1 & 0 & 0 \\ 0 & \lambda_2 & 0 \\ 0 & 0 & \lambda_3 \end{bmatrix} \tag{1.45}$$

The diagonal components are thus given directly by the roots of equation (1.39).

It remains to discuss the degenerate cases where two or more of the eigenvalues are equal. First consider the case where only two eigenvalues, say λ_1 and λ_2, are equal. In this case, it is easily shown that the tensor components T_{ij} can still be diagonalized with diagonal components $T'_{11} = T'_{22} = \lambda_1$ and $T'_{33} = \lambda_3$ provided only that the unit vectors i'_1, i'_2, i'_3 be chosen such that i'_3 lies along e_3 and i'_1 and i'_2 lie in any two mutually perpendicular directions.

To see this, first choose a new set of Cartesian axes x'_i having unit vectors i'_1, i'_2, i'_3 such that $i'_1 = e_1$, $i'_3 = e_3$, and $i'_2 = e_3 \times e_1$. In this case, we have from equation (1.36)

$$T'_{ij}e'_{1j} = \lambda_1 e'_{1i}, \qquad T'_{ij}e'_{3j} = \lambda_3 e'_{3i} \tag{1.46}$$

Remembering again that $e'_{1j} = \delta_{1j}$ and $e'_{3j} = \delta_{3j}$, we find from these equations that

$$T'_{11} = \lambda_1, \qquad T'_{21} = T'_{31} = 0 \tag{1.47}$$

and

$$T'_{33} = \lambda_3, \qquad T'_{13} = T'_{23} = 0 \tag{1.48}$$

Using these results, we also find that

$$T'_{ij}i'_{2j} = T'_{22}i'_{2i} \tag{1.49}$$

so that i_2 is an eigenvector having eigenvalue T'_{22}. Since $\lambda_3 = T'_{33}$ was chosen as the distinct eigenvalue, it thus follows from $\lambda_2 = \lambda_1$ that $T'_{22} = T'_{11}$. The tensor components in this system are therefore expressible as

$$[T'_{ij}] = \begin{bmatrix} \lambda_1 & 0 & 0 \\ 0 & \lambda_1 & 0 \\ 0 & 0 & \lambda_3 \end{bmatrix} \tag{1.50}$$

The above system of primed coordinates was chosen such that the x'_1 and x'_3 axes were along the distinct eigenvectors e_1 and e_3 and the x'_2

axis was mutually perpendicular. Using the tensor transformation relation of equation (1.25), it can, however, easily be shown that the form of equation (1.50) will remain unchanged under transformation to axes x_i'' which differ from the x_i' axes by an arbitrary rotation about the x_3' axis.

When only one eigenvalue is distinct, the direction associated with the corresponding distinct eigenvector can therefore be used with any two mutually perpendicular directions to form a system of principal axes.

We consider finally the case where no eigenvalues are distinct; that is, where all eigenvalues are equal, say, to λ_1. In this case, equation (1.39) is expressible simply as

$$(\lambda - \lambda_1)^3 = 0 \tag{1.51}$$

so that equation (1.36) requires the tensor components to be expressible as

$$[T_{ij}] = \begin{bmatrix} \lambda_1 & 0 & 0 \\ 0 & \lambda_1 & 0 \\ 0 & 0 & \lambda_1 \end{bmatrix} \tag{1.52}$$

Hence, the x_i axes are already principal axes. Moreover, by direct calculation, it can easily be seen that the transformation of these components to any new set of axes x_i' will leave the form of equation (1.52) unchanged. Thus, when all the eigenvalues are equal, any set of axes will serve as principal axes.

Example.   As an example of the above concepts, we determine the eigenvalues and eigenvectors of the symmetric tensor T_{ij} having components

$$[T_{ij}] = \begin{bmatrix} 2 & 0 & 1 \\ 0 & 0 & 0 \\ 1 & 0 & 0 \end{bmatrix}$$

For these components, equation (1.39) defining the eigenvalues becomes expressible as

$$\lambda^3 - 2\lambda^2 - \lambda = 0$$

The solutions are easily found as

$$\lambda_1 = 2.414, \qquad \lambda_2 = 0, \qquad \lambda_3 = -0.414$$

Taking $\lambda = \lambda_1$, the components e_{11}, e_{12}, e_{13} associated with the unit eigenvector $\mathbf{e}_1$ may be determined from any two of the relations given by equation (1.37) and the geometric relation requiring unit magnitude of e_1. Choosing the second and third expressions of equation (1.37) with $A_1 = e_{11}$, $A_2 = e_{12}$, $A_3 = e_{13}$, we thus have

$$-2.414e_{12} = 0$$
$$e_{11} - 2.414e_{13} = 0$$
$$e_{11}^2 + e_{12}^2 + e_{13}^2 = 1$$

The solution of these equations accordingly yields

$$e_{11} = \pm0.924, \qquad e_{12} = 0, \qquad e_{13} = \pm0.382$$

so that the unit eigenvector $\mathbf{e}_1$ is expressible as

$$\mathbf{e}_1 = \pm(0.924\mathbf{i}_1 + 0.382\mathbf{i}_3)$$

Similar considerations for the remaining two eigenvalues yield

$$\mathbf{e}_2 = \pm\mathbf{i}_2$$
$$\mathbf{e}_3 = \pm(-0.382\mathbf{i}_1 + 0.924\mathbf{i}_3)$$

Finally, on taking axes x_i' along the unit eigenvectors, we have the tensor components T_{ij}' represented as

$$[T_{ij}'] = \begin{bmatrix} 2.414 & 0 & 0 \\ 0 & 0 & 0 \\ 0 & 0 & -0.414 \end{bmatrix}$$

Selected Reading

Jeffreys, H., *Cartesian Tensors*. Cambridge University Press, England, 1931. This book is a standard reference on Cartesian tensors.

Long, R. R., *Mechanics of Solids and Fluids*. Prentice-Hall, Englewood Cliffs, New Jersey, 1961. Chapter 1 of this book gives a good introduction to Cartesian tensors.

Fung, Y. C., *Foundations of Solid Mechanics*. Prentice-Hall, Englewood Cliffs, New Jersey, 1965. Chapter 2 gives a brief introduction to tensors in generalized as well as Cartesian space.

Malvern, L. E., *Introduction to the Mechanics of a Continuous Medium*. Prentice-Hall, Englewood Cliffs, New Jersey, 1969. Chapter 2 and Appendix I of this book provide extensive discussion of tensors in both generalized and Cartesian space.

Exercises

1. Using simple geometry, show that the components A_1, A_2 and A_1', A_2' of a vector in two plane coordinate systems x_1, x_2 and x_1', x_2' are related by equations (1.5) and (1.6).

2. Transform the vector components $A_1 = \frac{1}{2}A$, $A_2 = (\sqrt{3}/2)A$, $A_3 = 0$, relative to axes x_i, to new axes x_i' which differ by a clockwise rotation of $30°$ about the x_3 axis.

3. Transform the tensor components

$$[T_{ij}] = \begin{bmatrix} 1 & 1 & 3 \\ 2 & 0 & 1 \\ 1 & 3 & 2 \end{bmatrix}$$

    relative to axes x_i to new axes x_i' which differ by a clockwise rotation of $60°$ about the x_1 axis.

4. By assuming complex eigenvalues and eigenvectors of the form

$$\lambda = b + ic, \qquad A_i = B_i + iC_i$$

    where $i = \sqrt{-1}$, show for symmetric tensor components T_{ij} that equation (1.36) requires that $c = 0$ and, hence, that the eigenvalues are all real.

5. Determine the eigenvalues and eigenvectors of the symmetric tensor T_{ij} having components relative to x_i axes given by

$$[T_{ij}] = \begin{bmatrix} 2 & 1 & 0 \\ 1 & 4 & 0 \\ 0 & 0 & 0 \end{bmatrix}$$

2

Kinematics of Continuum Motion

The mechanics of solids involves the motion of continuous, deformable solid material when acted on by applied forces. In the present chapter, we examine the geometry or *kinematics* of this motion without regard for the actual forces required to produce it. In particular, we associate a material point, or *particle*, with each geometric point of the material and take as our problem that of describing the geometry involved in the motion of these particles. Such considerations will lead us, in turn, to the well-known concepts of *deformation*, *strain*, and *rotation* in the neighborhood of a material point.

2.1. Material and Spatial Variables

The most obvious and direct method of describing the motion of a continuum solid is to consider the motion of each and every particle making up the solid. If X_i denotes the initial coordinates of any particle relative to a set of fixed Cartesian coordinate axes, its motion, as illustrated in Figure 2.1, may thus be represented by the general functions

$$x_i = x_i(X_j, t) \tag{2.1}$$

where x_i denotes the particle coordinates at time t.

In this description, the coordinates X_i are referred to as *material* or *Lagrangian variables* and the coordinates x_i are referred to as *spatial* or *Eulerian variables*.

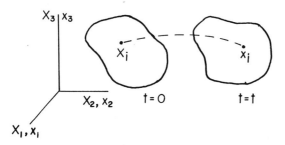

FIGURE 2.1.

In the discussion which follows, it is assumed that the determinant

$$J = \det \left| \frac{\partial x_i}{\partial X_j} \right| \tag{2.2}$$

is always positive, so that equation (2.1) may be inverted to give

$$X_i = X_i(x_j, t) \tag{2.3}$$

2.2. Definitions of Displacement, Velocity, and Acceleration

The *displacement* of a particle is defined as the vector difference between its current and initial positions. Thus, with u_i denoting the displacement components, we have

$$u_i = x_i - X_i \tag{2.4}$$

Clearly, u_i can be regarded as a function of either x_i or X_i by appropriate use of equation (2.1) or (2.3). In the former case, the displacement is said to be represented by a spatial description, and in the latter case, by a material description.

The *velocity* of a particle at any instant is likewise defined as the instantaneous rate of change of its vector position with respect to time. Thus, with v_i denoting the velocity components, we have from equation (2.1)

$$v_i = \frac{dx_i}{dt} = \frac{\partial x_i(X_j, t)}{\partial t} \tag{2.5}$$

where d/dt refers to differentiation holding the initial coordinates X_i fixed. Using equation (2.4), this relation is also seen to be equivalent to the relation

$$v_i = \frac{du_i}{dt} = \frac{\partial u_i(X_j, t)}{\partial t} \tag{2.6}$$

The *acceleration* of a particle at any instant is defined as the instantaneous rate of change of its velocity with respect to time. With a_i denoting the components of the acceleration, we thus have

$$a_i = \frac{dv_i}{dt} = \frac{\partial v_i(X_j, t)}{\partial t} \tag{2.7}$$

The above components of acceleration are seen to follow directly from the derivatives of the corresponding velocity components when these are expressed in terms of material variables X_i. It is, however, sometimes convenient to consider the velocity in terms of spatial variables rather than material variables by eliminating the material variables through the use of equation (2.3). When this is the case, we then have the velocity components v_i expressible as

$$v_i = v_i(x_j, t) \tag{2.8}$$

and their time derivatives, holding the initial particle coordinates fixed, are accordingly given by

$$\frac{dv_i}{dt} = \frac{\partial v_i}{\partial t} + \frac{\partial v_i}{\partial x_j} \frac{\partial x_j}{\partial t} \tag{2.9}$$

where the term $\partial x_j/\partial t$ is to be evaluated from equation (2.1) with the initial coordinates X_i held fixed. But from equation (2.5), this can easily be seen to be nothing more than the velocity components of the particle, so that the acceleration components in spatial variables are thus given by

$$a_i = \frac{dv_i}{dt} = \frac{\partial v_i}{\partial t} + v_j \frac{\partial v_i}{\partial x_j} \tag{2.10}$$

It is worth noting from the above argument that the operator d/dt is expressible, in general, in terms of spatial coordinates as

$$\frac{d}{dt} = \frac{\partial}{\partial t} + v_j \frac{\partial}{\partial x_j} \tag{2.11}$$

2.3. Deformation Gradients

Consider now two adjacent particles having initial coordinates X_i and X_i^*. At time t, these particles will have moved to new positions defined by x_i and x_i^* as shown in Figure 2.2.

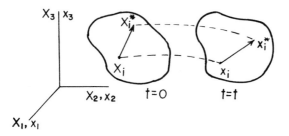

FIGURE 2.2.

Using equation (2.1), we can express the coordinates x_i^* at time t approximately in terms of x_i by a first-order Taylor series expansion of the form

$$x_i^* = x_i + \frac{\partial x_i}{\partial X_j} (X_j^* - X_j) \qquad (2.12)$$

where x_i and $\partial x_i/\partial X_j$ are evaluated at X_i and t. Taking dX_j to denote the components of the initial line element, as defined by $dX_j = X_j^* - X_j$, we can thus write the components dx_i of this line element at time t as

$$dx_i = F_{ij} \, dX_j \qquad (2.13)$$

where $dx_i = x_i^* - x_i$ and where F_{ij} denotes the *deformation gradients* given by

$$F_{ij} = \frac{\partial x_i}{\partial X_j} \qquad (2.14)$$

and evaluated at X_i and t.

From equation (2.13), it can easily be seen that the components of a given initial line element are dependent at any time t only on the current values of the deformation gradients. Also, by using the quotient rule of Chapter 1, it can be seen immediately from this same equation that the components F_{ij} are the components of a Cartesian tensor of second order.

It should be noted that the deformation gradients F_{ij} are, in general, dependent on the choice of the initial particle position X_i. When, however, the functions of equation (2.1) describing the motion are linear in their dependence on the initial coordinates, the deformation gradients will then have the same value for all material points. Such a case is referred to as a *homogeneous deformation*. Any other situation is likewise referred to as an *inhomogeneous deformation*.

2.4. Stretch and Angular Distortion of Line Elements

Consider next the length ds of the element shown in Figure 2.2 at time t. From the definition of the scalar product, we have immediately that

$$ds^2 = dx_i \, dx_i \tag{2.15}$$

But from equation (2.13), this may be written as

$$ds^2 = C_{mn} \, dX_m \, dX_n \tag{2.16}$$

where C_{mn} is given in terms of the deformation gradients by

$$C_{mn} = F_{im} F_{in} \tag{2.17}$$

Hence, with the *stretch* λ defined as the ratio ds/dS, where dS denotes the initial length of the element, we thus have from equation (2.16) that

$$\lambda = (C_{mn} N_m N_n)^{1/2} \tag{2.18}$$

where N_i denotes the direction cosines dX_i/dS of the initial line element. If we pick an initial line element defined at the point X_i by the direction cosines N_i, we may therefore calculate its stretch at time t directly from equation (2.18).

Now suppose two line elements dX_i and $d\tilde{X}_i$ emanate from the point X_i, as shown in Figure 2.3. The initial angle θ_0 between these two elements is given from the scalar product relation

$$dS \, d\tilde{S} \cos \theta_0 = dX_i \, d\tilde{X}_i \tag{2.19}$$

or

$$\cos \theta_0 = N_i \tilde{N}_i \tag{2.20}$$

where N_i and $\tilde{N}_i$ denote the corresponding initial direction cosines. To find

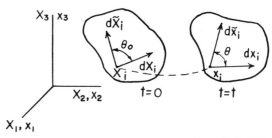

FIGURE 2.3.

the angle that these elements make with one another at time t, we may use the scalar product again to find

$$ds\ d\tilde{s} \cos \theta = dx_i\ d\tilde{x}_i \tag{2.21}$$

where ds and $d\tilde{s}$ denote the lengths of the elements at time t. Using equations (2.13) and (2.17), we thus have finally

$$\cos \theta = \frac{C_{mn} N_m \tilde{N}_n}{\lambda \tilde{\lambda}} \tag{2.22}$$

where λ and $\tilde{\lambda}$ denote the stretches of the line elements as given by equation (2.18).

2.5. Condition for Rigid-Body Motion of Material about a Point

An inspection of equations (2.18) and (2.22) will show that, when the *deformation tensor components* C_{mn} satisfy the equation

$$C_{mn} = \delta_{mn} \tag{2.23}$$

at some point, the length of a line element emanating from this point and the angle between two elements from the point will both be unchanged from their initial values. Equation (2.23) accordingly expresses the condition on the deformation such that the motion of material in the neighborhood of a point will, at most, be nothing more than a rigid translation and rigid rotation. In terms of the deformation gradients, we have from equations (2.17) and (2.23) that

$$F_{im} F_{in} = \delta_{mn} \tag{2.24}$$

Alternatively, if we interchange x_i and X_i axes in this equation and use the chain rule of differentiation, we may also find

$$F_{mi} F_{ni} = \delta_{mn} \tag{2.25}$$

The deformation gradients for this case are thus seen to satisfy the same orthogonality conditions as those determined in Chapter 1 for the direction-cosine array associated with coordinate transformations. We accordingly have the following important result:

When the deformation gradients are orthogonal at some point, that is, when they satisfy equations (2.24) and (2.25), the motion of neighboring material is, at most, nothing more than a rigid translation and rigid rotation.

2.6. Decomposition of Deformation Gradients

Consider again the deformation tensor components

$$C_{kj} = F_{ik}F_{ij} \tag{2.26}$$

as given by equation (2.17). Since the tensor defined by these components is clearly symmetric, we may express its components relative to principal axes following the procedure described in Chapter 1. Let C'_{kj} denote the components of C_{kj} when referred to these principal axes, and let U'_{kj} denote the principal components of a symmetric tensor defined by the positive square root of these components; that is, by

$$U'_{kj} = +(C'_{kj})^{1/2} \tag{2.27}$$

Also let G'_{kj} denote components of a symmetric tensor defined by

$$G'_{kj} = (U'_{kj})^{-1} \tag{2.28}$$

that is, by the inverse components of U'_{kj}. Finally, let R'_{ik} be the tensor components defined by

$$R'_{ik} = F'_{il}G'_{kl} \tag{2.29}$$

where F'_{il} denotes the deformation gradients when referred to the principal axes of C_{ij}. From this last equation and the definition of G'_{kl}, it is then easily verified that F'_{ij} and R'_{ik} satisfy the relations

$$F'_{ij} = R'_{ik}U'_{kj}, \qquad R'_{ik}R'_{il} = \delta_{kl} \tag{2.30}$$

in terms of the principal axes of C_{ij} and U_{ij}; or

$$F_{ij} = R_{ik}U_{kj}, \qquad R_{ik}R_{il} = \delta_{kl} \tag{2.31}$$

in terms of arbitrary axes.

If instead of using the tensor components C_{ij}, we consider the symmetric-tensor components B_{ij} defined by

$$B_{kj} = F_{ki}F_{ji} \tag{2.32}$$

we may employ reasoning analogous to that involved in obtaining equation (2.31) to find the relations

$$F_{ij} = V_{im}R_{mj}, \qquad R_{ki}R_{li} = \delta_{kl} \tag{2.33}$$

where the principal values of V_{im} in this case are equal to the positive square root of the principal values of B_{ij}.

Collecting together equations (2.31) and (2.33), we therefore have the complete result that

$$F_{ij} = R_{ik}U_{kj} = V_{ik}R_{kj} \tag{2.34}$$

where R_{ij} denotes components of an orthogonal tensor and U_{ij} and V_{ij} denote components of symmetric tensors. Equation (2.34) expresses the so-called *polar decomposition* of the deformation gradients. In this expression, the tensor components R_{ij} are those of the *rotation tensor* of the deformation and the components U_{ij} and V_{ij} are those, respectively, of the *right and left Cauchy–Green stretch tensors*. Also the components C_{ij} and B_{ij} given above are, respectively, those of the *right and left Cauchy–Green deformation tensors*.

2.7. General Motion of Material in the Neighborhood of a Point

With the above decomposition of the deformation gradients in mind, let us now consider in detail the general motion of translation, rigid rotation, and distortion of an initial line element dX_i emanating from the material point X_i, as shown in Figure 2.4. From equations (2.13) and (2.31), we have, in particular, that the components dx_i of this line element at time t can be expressed as

$$dx_i = R_{ik}U_{kj}\,dX_j \tag{2.35}$$

In order to gain further insight into the nature of this expression, it is convenient to consider separately the effects of U_{kj} operating on dX_j and R_{ik} operating on this resultant. For this purpose, it is especially convenient to choose the initial line element dX_j along one of the principal directions of

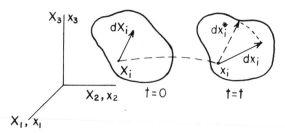

FIGURE 2.4.

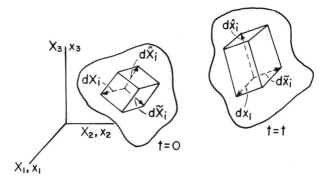

FIGURE 2.5.

U_{kj}; for in this case, dX_j will, by definition, be components of an eigenvector of U_{kj} and, hence, will satisfy

$$dx_k^* = U_{kj}\,dX_j = U_1\,dX_k \qquad (2.36)$$

where U_1 is the corresponding eigenvalue associated with the eigenvector defined by dX_k. This equation shows immediately that the operation of U_{kj} on dX_j causes nothing more than a stretch of each of its components by the amount U_1 and hence a stretch of the line element itself by the amount U_1.

Substituting the above relation into equation (2.35), we next find that

$$dx_i = R_{ik}\,dx_k^* \qquad (2.37)$$

Since R_{ik} is orthogonal, the considerations of Section 2.5 show finally that the operation of R_{ik} on the stretched element dx_k^* has the effect of simply rotating it as though it were a rigid element. The operations of translation and stretch are shown by the dashed arrow in Figure 2.4. The operations of translation, stretch, and rigid rotation are illustrated by the longer solid arrow.

A similar discussion is easily seen to apply to initial line elements lying along the remaining two principal directions of U_{kj}.

From the above results, it is thus seen that:

The most general motion of a small volume element of material having sides defined initially by line elements lying along the three mutually perpendicular principal directions of the right stretch tensor can be regarded as a translation of the volume element, a stretching along its edges, and finally a rotation of the stretched volume element as though it were a small rigid body. This motion is illustrated in Figure 2.5.

It remains to discuss the behavior of line elements not lying along the principal directions of U_{kj}. In this case, it is easily established that the operation $U_{kj} dX_j$ in equation (2.35) causes not only a stretching of the translated initial line element but also a rotation so that, according to this equation, the deformation of a general line element can thus be thought of as a stretch and rotation associated with the operation $U_{kj} dX_j$ together with an additional local rigid rotation associated with the operation of R_{ik} on the resultant.

To see this, we have only to consider the direction cosines n_k^* of the element given by

$$dx_k^* = U_{kj} dX_j \tag{2.38}$$

Remembering that $n_k^* = dx_k^*/ds^*$, where ds^* denotes the length of the element dx_k^*, we then have from equation (2.38) that

$$n_k^* = \frac{U_{kj} N_j}{\lambda^*} \tag{2.39}$$

where N_j denotes the direction cosines dX_j/dS of the initial line element dX_j having initial length dS, and where $\lambda^* = ds^*/dS$. This equation thus shows that the direction cosines of the element dx_k^* will generally differ from those of the element dX_j except when dX_j is an eigenvector of U_{kj}. This motion is illustrated in Figure 2.6, where the shorter dashed arrow shows the operation of translation, the longer dashed arrow shows translation and stretch, and the longer solid arrow shows translation, stretch, and rotation.

An immediate consequence of the above result is that: The motion of a small volume element of material having sides defined initially by line elements not along the principal directions of the right stretch tensor can, in general, be regarded as consisting of a translation of the element, a stretching and simultaneous angular distortion of its sides, and, finally, a rotation of

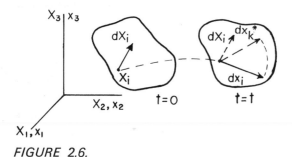

FIGURE 2.6.

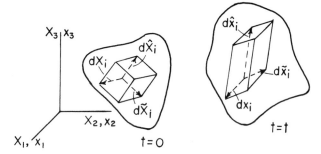

FIGURE 2.7.

this distorted element as though it were a small rigid body. This situation is illustrated in Figure 2.7.

In the above discussion concerning the motion of line elements, the process has been regarded from the point of view of equation (2.35) as consisting of a translation of the elements, a stretching, and, finally, a rotation as though the deformed elements were rigid. The process can, however, also be regarded alternatively from the point of view of equations (2.13) and (2.33), namely

$$dx_i = V_{ik}R_{kj}\, dX_j \tag{2.40}$$

When this equation is used in place of equation (2.35) in a discussion analogous to that given above, we may then regard the motion of a general volume element at a point as consisting of a translation of the element, a rotation as though it were a small rigid body, and, finally, a stretching and angular distortion of its sides.

Also, when the volume element has defining edges, which, after the rigid rotation, coincide with the principal directions of the stretch tensor defined by V_{ik}, the motion can then be thought of as a translation of the volume element, a rotation as though the volume element were rigid, and, finally, a stretching without angular distortion of its edges. The condition for this to be the case can be seen from equation (2.34) to be simply that the edges of the volume element coincide initially with the principal directions of the stretch tensor defined by U_{kj}, as discussed earlier.

2.8. Approximations Valid for Small Deformations

We now consider simplifications of the above formulas that are valid for small deformations. Small deformations are defined as those for which the *displacement gradients* are all small in comparison with unity. From equation (2.4), the displacement gradients may be determined as

$$\frac{\partial u_i}{\partial X_j} = \frac{\partial x_i}{\partial X_j} - \delta_{ij} \tag{2.41}$$

Hence, the concept of small deformations implies that

$$\frac{\partial u_i}{\partial X_j} \ll 1 \tag{2.42}$$

or, equivalently, that products of displacement gradients can be neglected in comparison with first-order terms.

It is an easy matter to show, using equation (2.41), that the approximation of equation (2.42) implies for any differentiable function f that

$$\frac{\partial f}{\partial X_i} = \frac{\partial f}{\partial x_i} \tag{2.43}$$

so that derivatives with respect to initial coordinates are indistinguishable from derivatives with respect to the spatial coordinates.

2.9. Motion in the Neighborhood of a Point for Small Deformations

We consider again the decomposition of the deformation gradients F_{ij} according to the first part of equation (2.34), namely

$$F_{ij} = R_{ik}U_{kj} \tag{2.44}$$

For the case of small deformations, the determination of R_{ik} and U_{kj} is much simpler than that considered earlier for general deformations. In particular, with the tensor components e_{ij} and w_{ij} defined in terms of the displacement gradients by the expressions

$$e_{ij} = \frac{1}{2}\left(\frac{\partial u_i}{\partial X_j} + \frac{\partial u_j}{\partial X_i}\right) \tag{2.45}$$

$$w_{ij} = \frac{1}{2}\left(\frac{\partial u_i}{\partial X_j} - \frac{\partial u_j}{\partial X_i}\right) \tag{2.46}$$

we find, in fact, that the rotation and stretch tensor components R_{ik} and U_{kj}

are expressible for small deformations simply as

$$R_{ik} = \delta_{ik} + w_{ik} \tag{2.47}$$

$$U_{kj} = \delta_{kj} + e_{kj} \tag{2.48}$$

To see that these tensors do indeed provide a solution to equation (2.44), we have only to substitute them into this equation and retain only first-order terms involving e_{ij} and w_{ij}.

In a similar way, it may also be seen that with $V_{ik} = U_{ik}$ equations (2.47) and (2.48) likewise yield first-order solutions to the decomposition expressed by the second part of equation (2.34), namely

$$F_{ij} = V_{ik} R_{kj} \tag{2.49}$$

Thus, for small deformations, there is no difference between the stretch-tensor components U_{ij} and V_{ij}.

The tensors having components e_{ij} and w_{ij} as defined by equations (2.45) and (2.46) are referred to as the *strain and rotation tensors of small deformation*.

Using equation (2.13) and the above decomposition, we may express the deformation of an initial line element dX_i as

$$dx_i = dX_i + (e_{ij} + w_{ij}) \, dX_j \tag{2.50}$$

Noting that $w_{ij} = -w_{ji}$, we may use the results of equation (1.34) of Chapter 1 for antisymmetric tensors to write

$$w_{ij} \, dX_j = e_{ijk} w_j \, dX_k \tag{2.51}$$

where w_j denotes components of the *rotation vector* as defined by

$$w_1 = w_{32}, \qquad w_2 = w_{13}, \qquad w_3 = w_{21} \tag{2.52}$$

Thus, combining equations (2.50) and (2.51) and noting that $dx_i - dX_i$ denotes the displacement components du_i, we find

$$du_i = e_{ij} \, dX_j + e_{ijk} w_j \, dX_k \tag{2.53}$$

which shows that the *relative displacement* of two adjacent particles is equal to the sum of two contributions, one arising from the strain tensor e_{ij} and the other arising from the rotation vector w_i.

It is to be carefully noted that equation (2.53) only applies for small deformations and that for more general deformations, equation (2.34) must be employed.

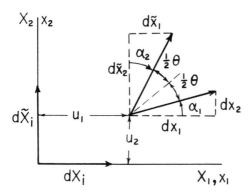

FIGURE 2.8.

2.10. Geometric Interpretation of Strain and Rotation Components of Small Deformation

The above equations defining the strain and rotation tensor components for small deformation are subject to simple geometric interpretation. Consider, for example, two initial line elements dX_i and $d\tilde{X}_i$ emanating from some point X_i and having initial directions lying along the X_1 and X_2 axes, respectively. After deformation, these elements will, in general, be displaced, stretched, and rotated. Moreover, for small deformation, these deformed elements will be indistinguishable from their projections onto the X_1 vs. X_2 plane so that the entire process can be represented in two dimensions, as shown in Figure 2.8.

From equation (2.13) and the definition of displacement, the components dx_i after deformation of an initial line element dX_i are easily seen to be expressible as

$$dx_i = \frac{\partial x_i}{\partial X_j} dX_j = \left(\delta_{ij} + \frac{\partial u_i}{\partial X_j} \right) dX_j \tag{2.54}$$

For the above line element dX_i lying initially in the X_1 direction, we thus have its components after deformation given by

$$dx_1 = \left(1 + \frac{\partial u_1}{\partial X_1} \right) dX_1$$

$$dx_2 = \frac{\partial u_2}{\partial X_1} dX_1 \tag{2.55}$$

$$dx_3 = \frac{\partial u_3}{\partial X_1} dX_1$$

Similarly, for the element $d\tilde{X}_i$ lying initially in the X_2 direction, we have

$$d\tilde{x}_1 = \frac{\partial u_1}{\partial X_2}\,d\tilde{X}_2$$

$$d\tilde{x}_2 = \left(1 + \frac{\partial u_2}{\partial X_2}\right)d\tilde{X}_2 \qquad (2.56)$$

$$d\tilde{x}_3 = \frac{\partial u_3}{\partial X_2}\,d\tilde{X}_2$$

Normal Strain Components.   Consider now the length of the element dX_1 after deformation. From geometry, this length ds is expressible as

$$ds = (dx_1^2 + dx_2^2 + dx_3^2)^{1/2} \qquad (2.57)$$

Substituting the components given by equation (2.55) into this expression and assuming the displacement gradients small in comparison with unity, we find

$$ds = dX_1 + \frac{\partial u_1}{\partial X_1}\,dX_1 \qquad (2.58)$$

On rearranging and using the definition of the strain component e_{11}, we thus have the equation

$$e_{11} = \frac{\partial u_1}{\partial X_1} = \frac{ds - dX_1}{dX_1} \qquad (2.59)$$

which shows that the strain component e_{11} represents simply the change in length per unit of initial length, or *unit extension*, of a line element lying initially in the X_1 direction. In a similar way, it can also be seen that the small strain components e_{22} and e_{33} likewise represent unit extensions of line elements lying initially in the X_2 and X_3 directions, respectively. Because of this interpretation, the strain components e_{11}, e_{22}, and e_{33} are referred to as normal strain components.

Shearing Strain Components.   From equation (2.55), it can be seen that the angle α_1 between the line element dx_i and its initial X_1 direction is expressible for small deformation as

$$\alpha_1 \approx \tan \alpha_1 = \frac{\partial u_2}{\partial X_1} \qquad (2.60)$$

In a like manner, we may also see from equation (2.56) that the angle α_2

between the element $d\tilde{x}_i$ and its initial X_2 direction is expressible for small deformation as

$$\alpha_2 \approx \tan \alpha_2 = \frac{\partial u_1}{\partial X_2} \qquad (2.61)$$

Thus, from the definition of the strain component e_{21}, we have

$$e_{21} = \frac{1}{2}\left(\frac{\partial u_2}{\partial X_1} + \frac{\partial u_1}{\partial X_2}\right) = \frac{1}{2}(\alpha_1 + \alpha_2) \qquad (2.62)$$

so that the component e_{21} is seen to represent simply one-half the change in the angle between line elements lying initially in the X_1 and X_2 directions. A similar interpretation also applies to the strain components e_{23} and e_{13} so that these components thus represent one-half the change in the angle between line elements originally lying in the X_2 and X_3 directions and in the X_1 and X_3 directions, respectively. For these reasons, the strain components $e_{12} = e_{21}$, $e_{23} = e_{32}$, and $e_{13} = e_{31}$ are referred to as shearing strain components.

Rotation Components. If $\theta = \pi/2 - \alpha_1 - \alpha_2$ denotes the angle between the above two line elements dx_i and $d\tilde{x}_i$ after deformation, the angle between the bisector of θ and the X_1 axis is seen from Figure 2.8 to be given by

$$\tfrac{1}{2}\theta + \alpha_1 = \tfrac{1}{4}\pi + \tfrac{1}{2}(\alpha_1 - \alpha_2) \qquad (2.63)$$

Since the angle between the bisector of the initial elements and the X_1 axis is $\pi/4$, the counterclockwise rotation of the bisector resulting from the deformation is expressible as

$$\tfrac{1}{2}\theta + \alpha_1 - \tfrac{1}{4}\pi = \tfrac{1}{2}(\alpha_1 - \alpha_2) \qquad (2.64)$$

But this is nothing more than the rotation component $w_3 = w_{21}$ as given by

$$w_3 = \frac{1}{2}\left(\frac{\partial u_2}{\partial X_1} - \frac{\partial u_1}{\partial X_2}\right) \qquad (2.65)$$

so that the rotation component w_3 thus represents simply the counterclockwise rotation of the bisector of the angle between two line elements initially pointing in the X_1 and X_2 directions. Similarly, the rotation components $w_1 = w_{32}$ and $w_2 = w_{13}$ represent rotations of the bisectors of line elements initially lying in the X_2 and X_3 and in the X_1 and X_3 directions, respectively.

Dilatation. Using the above interpretations, we may also give geometric meaning to the sum of the normal strains, e_{ii}. Consider, in particular, a small volume element about some point having edges parallel initially to the three coordinate axes. If d_1, d_2, and d_3 denote the initial lengths of of these edges, the initial volume of the element will be given by $d_1 d_2 d_3$ and its volume after a small general strain will be given by $d_1(1 + e_{11})d_2(1 + e_{22})d_3(1 + e_{33})$. Hence, the change in volume per unit of initial volume, Δ, for small strain is expressible as

$$\Delta = e_{11} + e_{22} + e_{33} \tag{2.66}$$

This quantity is referred to as the dilatation.

2.11. Examples of Small Deformation

Pure Dilatation. Consider the deformation defined by

$$
\begin{aligned}
x_1 &= X_1 + \varepsilon X_1 \\
x_2 &= X_2 + \varepsilon X_2 \\
x_3 &= X_3 + \varepsilon X_3
\end{aligned}
\tag{2.67}
$$

where ε denotes a constant small in comparison with unity. The displacement components are clearly

$$u_1 = \varepsilon X_1, \qquad u_2 = \varepsilon X_2, \qquad u_3 = \varepsilon X_3 \tag{2.68}$$

and the only nonzero components of the strain and rotation tensors are seen from equations (2.45) and (2.46) to be given by

$$e_{11} = e_{22} = e_{33} = \varepsilon \tag{2.69}$$

From our previous geometric interpretation of the normal strain components, we see that this deformation involves simply an equal extension of line elements emanating from some initial point and lying initially in the X_1, X_2, and X_3 directions. Also, since the deformation is irrotational, the direction of the deformed elements will be unchanged from their initial direction. Thus, a small cube of material about some initial point having its edges parallel to the coordinate axes will become simply a larger cube after the deformation with no change in its orientation (Figure 2.9). From the previous geometric interpretation we also have that the change in volume per unit initial volume of this cube is

$$e_{ii} = 3\varepsilon \tag{2.70}$$

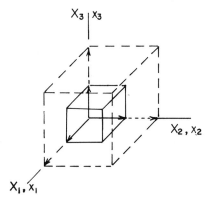

FIGURE 2.9.

Plane Deformation. Consider next the deformation defined by

$$x_1 = (1 + \varepsilon_1)X_1 + 2\gamma_1 X_2$$
$$x_2 = 2\gamma_2 X_1 + (1 + \varepsilon_2)X_2 \qquad (2.71)$$
$$x_3 = X_3$$

where ε_1, ε_2, γ_1, and γ_2 denote constants which are small in comparison with unity. The displacement components are

$$u_1 = \varepsilon_1 X_1 + 2\gamma_1 X_2$$
$$u_2 = 2\gamma_2 X_1 + \varepsilon_2 X_2 \qquad (2.72)$$
$$u_3 = 0$$

and the strain- and rotation-tensor components are given from equations (2.45) and (2.46) as

$$[e_{ij}] = \begin{bmatrix} \varepsilon_1 & \gamma_1 + \gamma_2 & 0 \\ \gamma_1 + \gamma_2 & \varepsilon_2 & 0 \\ 0 & 0 & 0 \end{bmatrix} \qquad (2.73)$$

$$[w_{ij}] = \begin{bmatrix} 0 & \gamma_1 - \gamma_2 & 0 \\ \gamma_2 - \gamma_1 & 0 & 0 \\ 0 & 0 & 0 \end{bmatrix} \qquad (2.74)$$

The rotation-vector components are also determined from equation (2.52) as

$$w_1 = w_2 = 0, \qquad w_3 = \gamma_2 - \gamma_1 \qquad (2.75)$$

From our earlier geometric interpretation, we thus see that the deformation

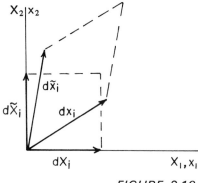

FIGURE 2.10.

takes place entirely in the X_1 vs. X_2 plane and can be regarded as extensions, relative angular distortion, and rotations of line elements emanating from some initial point and having initial directions along the X_1 and X_2 axes (Figure 2.10).

When $\gamma_1 = \varepsilon_2 = \gamma_2 = 0$ in the above equations, the deformation is referred to as a *simple extension* in the X_1 direction. Similarly, when $\varepsilon_1 = \varepsilon_2 = \gamma_2 = 0$, the deformation is referred to as a *simple shear* parallel to the X_1 axis. Finally, when $\varepsilon_1 = \varepsilon_2 = 0$ and $\gamma_1 = \gamma_2$, the deformation is then referred to as a *pure shear* in the X_1 vs. X_2 plane.

It is instructive to consider the transformation of the above strain and rotation components to new axes X_1', X_2' inclined with respect to the X_1, X_2 axes by the angle θ as shown in Figure 2.11. From equation (1.25) of Chapter 1, we have the strain- and rotation-tensor components e_{ij} and w_{ij} relative to the primed axes expressible as

$$e'_{ij} = a_{im}a_{jn}e_{mn}$$

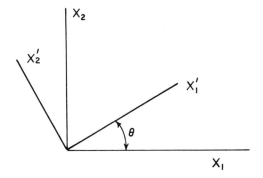

FIGURE 2.11.

and

$$w'_{ij} = a_{im}a_{jn}w_{mn}$$

where a_{ij} denotes the direction cosine between the X'_i and X_j axes. Referring to Figure 2.11, we find

$$[a_{ij}] = \begin{bmatrix} \cos\theta & \sin\theta & 0 \\ -\sin\theta & \cos\theta & 0 \\ 0 & 0 & 1 \end{bmatrix}$$

so that e'_{11}, e'_{22}, e'_{12}, and $w'_{21} = w'_3$ are expressible as

$$
\begin{aligned}
e'_{11} &= e_{11}\cos^2\theta + e_{22}\sin^2\theta + 2e_{12}\sin\theta\cos\theta \\
e'_{22} &= e_{11}\sin^2\theta + e_{22}\cos^2\theta - 2e_{12}\sin\theta\cos\theta \\
e'_{12} &= (e_{22} - e_{11})\sin\theta\cos\theta + e_{12}(\cos^2\theta - \sin^2\theta) \\
w'_{21} &= w_{21}
\end{aligned}
\tag{2.76}
$$

If we consider two line elements emanating from some initial point and lying in the X'_1 and X'_2 directions, we may apply the earlier geometric interpretation to see that e'_{11} and e'_{22} represent unit extensions of these elements and that e'_{12} represents one-half the change in angle between them. The rotation component $w'_{21} = w'_3$ likewise represents the rotation of the bisector of the angle between these elements.

To find the angle θ defining *principal axes of strain* for this case of plane deformation, we have only to set $e'_{12} = 0$ in the above equation and solve for θ. This gives

$$\tan 2\theta = \frac{2e_{12}}{e_{11} - e_{22}} \tag{2.77}$$

Substituting this value of θ into the expressions for e_{11} and e_{22}, we then find after some reduction that the *principal strain components* are given as

$$
\begin{aligned}
e'_{11} &= \frac{e_{11} + e_{22}}{2} + \frac{1}{2}\left[(e_{11} - e_{22})^2 + 4e_{12}^2\right]^{1/2} \\
e'_{22} &= \frac{e_{11} + e_{22}}{2} - \frac{1}{2}\left[(e_{11} - e_{22})^2 + 4e_{12}^2\right]^{1/2}
\end{aligned}
\tag{2.78}
$$

2.12. Unabridged Notation

An inspection of the equations developed in the present chapter will clearly show the value of the tensor notation in presenting these concepts in a concise fashion. When applying these relations to a specific problem, it is, however, often convenient to use instead a more conventional means of notation.

In particular, with this unabridged notation, we customarily take initial coordinates x, y, and z as

$$x = X_1, \qquad y = X_2, \qquad z = X_3 \tag{2.79}$$

and displacement components u, v, w as

$$u = u_1, \qquad v = u_2, \qquad w = u_3 \tag{2.80}$$

The small strain components $e_{xx} = e_{11}$, $e_{xy} = e_{12}$, etc. of equation (2.45) are correspondingly taken as

$$
\begin{aligned}
e_{xx} &= \frac{\partial u}{\partial x}, & e_{xy} &= \frac{1}{2}\left(\frac{\partial u}{\partial y} + \frac{\partial v}{\partial x}\right) \\
e_{yy} &= \frac{\partial v}{\partial y}, & e_{yz} &= \frac{1}{2}\left(\frac{\partial v}{\partial z} + \frac{\partial w}{\partial y}\right) \\
e_{zz} &= \frac{\partial w}{\partial z}, & e_{xz} &= \frac{1}{2}\left(\frac{\partial u}{\partial z} + \frac{\partial w}{\partial x}\right)
\end{aligned}
\tag{2.81}
$$

and the components of the small rotation vector $w_x = w_1$, etc., of equation (2.52) are taken as

$$
\begin{aligned}
w_x &= \frac{1}{2}\left(\frac{\partial w}{\partial y} - \frac{\partial v}{\partial z}\right) \\
w_y &= \frac{1}{2}\left(\frac{\partial u}{\partial z} - \frac{\partial w}{\partial x}\right) \\
w_z &= \frac{1}{2}\left(\frac{\partial v}{\partial x} - \frac{\partial u}{\partial y}\right)
\end{aligned}
\tag{2.82}
$$

2.13. Cylindrical Polar Coordinates

For future reference in discussing the deformation of solids having circular boundaries, we now wish to express the above equations for small deformation in terms of cylindrical polar coordinates. These coordinates are illustrated in Figure 2.12. From this figure, it is easily seen that the coordinates (x, y, z) of a point P are expressible in terms of the cylindrical polar coordinates (r, θ, z) by the equations

$$x = r \cos \theta, \qquad y = r \sin \theta, \qquad z = z \tag{2.83}$$

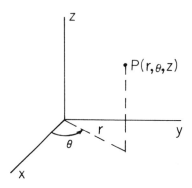

FIGURE 2.12.

To define the displacement, strain, and rotation components in this set of polar coordinates, we choose a new set of rectangular axes x', y', z' with z' along z and x' and y' along the directions of increasing r and θ, as shown in Figure 2.13. With reference to the x', y', z' axes of Figure 2.13, we now define the polar components of displacement as follows:

$$u_r = u', \qquad u_\theta = v', \qquad u_z = w' \tag{2.84}$$

where u', v', and w' denote the rectangular displacement components when referred to the x', y', z' axes.

Similarly, the strain and rotation components in polar coordinates are defined as

$$\begin{aligned} e_{rr} &= e'_{xx}, & e_{r\theta} &= e'_{xy}, & e_{rz} &= e'_{xz} \\ e_{\theta\theta} &= e'_{yy}, & e_{\theta z} &= e'_{yz}, & e_{zz} &= e'_{zz} \end{aligned} \tag{2.85}$$

and

$$w_r = w'_x, \qquad w_\theta = w'_y, \qquad w_z = w'_z \tag{2.86}$$

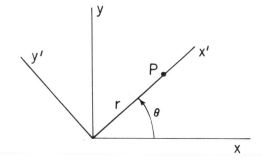

FIGURE 2.13.

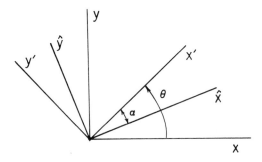

FIGURE 2.14.

where e'_{xx}, e'_{xy}, etc., denote the strain components referred to the rectangular axes x', y', z' and where w'_x, etc., denote the vector rotation components referred to these axes.

It remains to relate the above strain and rotation components to the displacement components u_r, u_θ and u_z. For this purpose, consider a set of rectangular axes $\hat{x}$, $\hat{y}$, $\hat{z}$ differing from the x', y', z' axes by the angle α from the x' axis as shown in Figure 2.14. The coordinates $(\hat{x}, \hat{y}, \hat{z})$ associated with these axes are seen from this figure to be expressible in terms of the polar coordinates (r, α, z) as

$$\hat{x} = r \cos \alpha, \qquad \hat{y} = r \sin \alpha, \qquad \hat{z} = z \qquad (2.87)$$

From these relations, we can easily establish the operators

$$\frac{\partial}{\partial \hat{x}} = \cos \alpha \, \frac{\partial}{\partial r} - \frac{\sin \alpha}{r} \, \frac{\partial}{\partial \alpha}$$

$$\frac{\partial}{\partial \hat{y}} = \sin \alpha \, \frac{\partial}{\partial r} + \frac{\cos \alpha}{r} \, \frac{\partial}{\partial \alpha} \qquad (2.88)$$

$$\frac{\partial}{\partial \hat{z}} = \frac{\partial}{\partial z}$$

Also, the displacement components $\hat{u}$, $\hat{v}$, and $\hat{w}$ are related to u_r, u_θ, and u_z through the vector transformation relations

$$\hat{u} = u_r \cos \alpha - u_\theta \sin \alpha$$

$$\hat{v} = u_r \sin \alpha + u_\theta \cos \alpha \qquad (2.89)$$

$$\hat{w} = u_z$$

Hence, on forming $\partial\hat{u}/\partial\hat{x}$ with the above relations, we have

$$\frac{\partial\hat{u}}{\partial\hat{x}} = \cos^2\alpha\,\frac{\partial u_r}{\partial r} + \frac{\sin^2\alpha}{r}\left(u_r + \frac{\partial u_\theta}{\partial\alpha}\right)$$

$$+ (\sin\alpha\cos\alpha)\left(\frac{u_\theta}{r} - \frac{1}{r}\frac{\partial u_r}{\partial\alpha} - \frac{\partial u_\theta}{\partial r}\right) \qquad (2.90)$$

Remembering that $\hat{e}_{xx} = \partial\hat{u}/\partial\hat{x}$ and noticing that $\hat{e}_{xx} = e'_{xx} = e_{rr}$ when $\alpha = 0$, we thus find from this equation, with $\alpha = 0$, the relation

$$e_{rr} = \frac{\partial u_r}{\partial r} \qquad (2.91)$$

In a similar way, we may also determine corresponding relations for the remaining strain components. In cylindrical polar coordinates, we may thus establish the strain-displacement relations as

$$e_{rr} = \frac{\partial u_r}{\partial r}$$

$$e_{\theta\theta} = \frac{1}{r}\frac{\partial u_\theta}{\partial\theta} + \frac{u_r}{r}$$

$$e_{zz} = \frac{\partial u_z}{\partial z}$$

$$2e_{r\theta} = \frac{1}{r}\frac{\partial u_r}{\partial\theta} + \frac{\partial u_\theta}{\partial r} - \frac{u_\theta}{r} \qquad (2.92)$$

$$2e_{\theta z} = \frac{\partial u_\theta}{\partial z} + \frac{1}{r}\frac{\partial u_z}{\partial\theta}$$

$$2e_{rz} = \frac{\partial u_r}{\partial z} + \frac{\partial u_z}{\partial r}$$

The rotation-displacement relations are similarly determined as

$$2w_r = \frac{1}{r}\frac{\partial u_z}{\partial\theta} - \frac{\partial u_\theta}{\partial z}$$

$$2w_\theta = \frac{\partial u_r}{\partial z} - \frac{\partial u_z}{\partial r} \qquad (2.93)$$

$$2w_z = \frac{1}{r}\frac{\partial(ru_\theta)}{\partial r} - \frac{1}{r}\frac{\partial u_r}{\partial\theta}$$

Selected Reading

Truesdell, C., and R. A. Toupin, "The Classical Field Theories," in *Encyclopedia of Physics*, Vol. III/1. Springer-Verlag, Berlin, 1960. A monumental treatise on continuum theory, including detailed historical remarks. Motion and deformation are discussed in Chapter B.

Eringen, A. C., *Nonlinear Theory of Continuous Media*. McGraw-Hill Book Co., New York, 1962. An advanced treatment of continuum mechanics. Chapters 1 and 2 provide a detailed discussion of the kinematics of continuum motion.

Frederick, D., and T. S. Chang, *Continuum Mechanics*. Allyn and Bacon, Boston, Massachusetts, 1965. A readable introductory text on continuum mechanics. Chapter 3 gives a discussion of the analysis of deformation.

Malvern, L. E., *Introduction to the Mechanics of a Continuous Medium*. Prentice-Hall, Englewood Cliffs, New Jersey, 1969. Chapter 4 gives a thorough discussion of strain and deformation using Cartesian tensor notation.

Exercises

1. Decompose the motion

$$x_1 = 2X_1, \qquad x_2 = -3X_3, \qquad x_3 = 2X_2$$

into rotation and stretch parts according to the first form of equation (2.34), namely

$$F_{ij} = R_{ik}U_{kj}$$

2. Using equation (2.34), show that the stretch components V_{ip} are expressible as

$$V_{ip} = R_{pj}R_{ik}U_{kj}$$

With this result, determine the components V_{ip} of the motion of Exercise 1.

3. Show that the right Cauchy–Green deformation tensor C_{ij} can be expressed without approximation as

$$C_{ij} = \delta_{ij} + 2L_{ij}$$

where

$$L_{ij} = \frac{1}{2}\left(\frac{\partial u_i}{\partial X_j} + \frac{\partial u_j}{\partial X_i} + \frac{\partial u_k}{\partial X_i}\,\frac{\partial u_k}{\partial X_j}\right)$$

denotes the *Lagrangian finite strain tensor*.

4. Show that the tensor defined by

$$c_{ij} = \frac{\partial X_i}{\partial x_j}$$

is the *inverse* of the left Cauchy–Green deformation tensor B_{ij}; that is, show

$$B_{ik}c_{kj} = \delta_{ij}$$

5. Show that c_{ij} as defined in Exercise 4 can be expressed as

$$c_{ij} = \delta_{ij} - 2E_{ij}$$

where

$$E_{ij} = \frac{1}{2}\left(\frac{\partial u_i}{\partial x_j} + \frac{\partial u_j}{\partial x_i} - \frac{\partial u_k}{\partial x_i}\frac{\partial u_k}{\partial x_j}\right)$$

denotes the *Eulerian finite strain tensor*.

6. Using the results of Section 2.4, show for small deformations that the unit extension e (or change in length per unit length) of a line element originally having direction cosines N_i is given by

$$e = e_{mn}N_mN_n$$

where e_{mn} denotes the strain components of small deformation.

7. If e_a, e_b, e_c denote unit extensions measured at a point on a surface $45°$, $90°$, and $135°$, respectively, from the x_1 axis, determine the strain components e_{xx}, e_{xy}, e_{yy} using the results of Exercise 6.

3

Governing Equations
of Motion

In the previous chapter, we considered certain geometric concepts associated with the motion of a continuous medium. We now wish to consider the set of mechanical laws that are customarily assumed to govern the actual continuum motion. These laws express the well-known mechanical principles of conservation of mass, balance of linear momentum, and balance of angular momentum.

Following the procedure of the last chapter, we shall first consider the equations representing these laws in their general nonlinear form and then later develop linear approximations applicable for the case of small deformations.

3.1. Conservation of Mass

The mass of a material particle is defined in classical mechanics as a measure of the amount of matter present within it. The principle of conservation of mass accordingly expresses the concept that the matter making up a particle, or collection of particles, can neither be increased nor decreased during any physically acceptable motion.

To establish the mathematical form of this principle for continuum bodies, we first define the *mass density* ϱ at time t and place x_i as

$$\varrho = \lim_{\Delta v \to 0} \frac{\Delta m}{\Delta v} = \frac{dm}{dv} \tag{3.1}$$

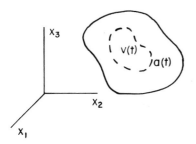

FIGURE 3.1.

where Δm denotes the increment of mass, or matter, contained within the volume element Δv located at the point x_i at time t.

We now consider an arbitrarily selected portion of a continuum mass at time t. The volume occupied by this mass at time t is denoted by $v(t)$ and the associated surface by $a(t)$, as shown in Figure 3.1.

The mass contained within this volume at time t is clearly given by the integral of the density $\varrho(x_i, t)$ over the volume $v(t)$. The principle of mass conservation thus requires that

$$\frac{d}{dt} \int_{v(t)} \varrho \, dv = 0 \qquad (3.2)$$

which is nothing more than the statement that the rate of change of the mass of any arbitrarily selected portion of the continuum must equal zero.

3.2. Balance of Linear Momentum

The linear momentum of a material particle is defined in classical mechanics as the product of its mass and velocity. The principle of the balance of linear momentum accordingly expresses the Newtonian concept that the time rate of change of the linear momentum of a particle or collection of particles, as measured from a fixed reference system, must equal the sum of the external forces acting on it.

To establish the mathematical form of this principle for a continuum body, we consider again the arbitrarily selected portion of material shown in Figure 3.1. We assume that at each point on the boundary of this continuum mass, there exists a contact force exerted by the material external to it. We denote the components of this force by ΔF_i and the components of the unit outward normal vector to the element of area Δa by n_i, as shown in Figure 3.2. The components of the *Cauchy stress vector* $t_i = t_i(x_j, t, n_j)$

are defined in terms of these quantities by the equation

$$t_i = \lim_{\varDelta a \to 0} \frac{\varDelta F_i}{\varDelta a} = \frac{dF_i}{da} \tag{3.3}$$

In addition to the above contact forces acting on the surface of the continuum mass, we also assume the existence of forces acting on its interior portions. We denote the components of this force at an interior point by $\varDelta F_i^*$ and define the components of the *body force vector* $f_i = f_i(x_j, t)$ by

$$f_i = \lim_{\varDelta v \to 0} \frac{\varDelta F_i^*}{\varDelta v} = \frac{dF_i^*}{dv} \tag{3.4}$$

where $\varDelta v$ denotes the volume element on which $\varDelta F_i^*$ acts, as shown in Figure 3.2.

With the above contact and body forces in mind, we may now consider the principle of linear momentum balance as applied to the arbitrarily selected continuum mass. Since the linear momentum of the mass is given by the integral over the volume v of the product of the density and velocity, we have, in particular, that

$$\frac{d}{dt} \int_{v(t)} \varrho v_i \, dv = \oint_{a(t)} t_i \, da + \int_{v(t)} f_i \, dv \tag{3.5}$$

This equation expresses the requirement that the time rate of change of the linear momenum of an arbitrarily selected piece of continuum mass equal the sum of the external forces acting on it. If we think of the mass as consisting of a collection of material particles, we may regard the equation as expressing a balance between the rate of change of their combined linear momenta and the external forces acting on them. For discrete particles an

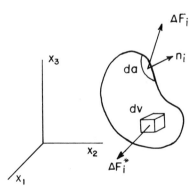

FIGURE 3.2.

equation of this kind with summations replacing integrals is, of course, derived in elementary mechanics from the direct application of Newton's second and third laws of particle motion.

3.3. Balance of Angular Momentum

The angular momentum of a particle is defined about some point as the vector product of its position vector and its linear momentum, the position vector being taken from the point about which the angular momentum is reckoned. The torque, or moment, of a force acting on the particle is likewise defined as the vector product of the position vector and the force. The principle of the balance of angular momentum accordingly expresses the Newtonian concept that the time rate of change of the angular momentum of a particle, or collection of particles, about a fixed point must equal the corresponding sum of all the external torques acting on it.

To establish the mathematical form of this principle for a continuum mass, we consider once again the arbitrarily selected piece of continuum mass shown in Figure 3.2 and require that the time rate of change of its entire angular momentum be equal to the sum of the external torques acting on it. Since the angular momentum at time t for the piece of continuum mass is given by the integral over the volume v of the cross product of the position vector to a point in the continuum and the linear momentum of the particle at that point, we have

$$\frac{d}{dt} \int_{v(t)} e_{ijk} x_j \varrho v_k \, dv = \oint_{a(t)} e_{ijk} x_j t_k \, da + \int_{v(t)} e_{ijk} x_j f_k \, dv \qquad (3.6)$$

where x_i denotes the components of the position vector.

Again, if the mass is regarded as a collection of material particles, this equation may be interpreted as a balance between the combined rate of change of the angular momenta of these particles and the sum of the external torques acting on them. For discrete particles, an equation of this kind with summations replacing integrals is likewise derived in elementary mechanics from the direct application of Newton's second and third laws of particle motion.

3.4. Evaluation of Time Derivative of Volume Integral

In order to express the above mass conservation and momentum balance laws in more useful forms, it is necessary to evaluate the time derivatives of the volume integrals appearing in them. These volume integrals are all of the form

$$I = \int_{v(t)} B(x_j, t)\, dv \tag{3.7}$$

where B denotes either a scalar or a scalar component of a vector. If we consider the change ΔI in this typical integral in time Δt, we have

$$\Delta I = \int_{v(t')} B(x_j, t')\, dv - \int_{v(t)} B(x_j, t)\, dv \tag{3.8}$$

where $t' = t + \Delta t$. Noticing that

$$B(x_i, t') = B(x_i, t) + \frac{\partial B}{\partial t}\, \Delta t$$

and

$$v(t') = v(t) + \Delta v$$

we can write this expression as

$$\Delta I = \int_{v(t)} \frac{\partial B}{\partial t}\, \Delta t\, dv + \int_{\Delta v} \left(B + \frac{\partial B}{\partial t}\, \Delta t \right) dv \tag{3.9}$$

From Figure 3.3, it can be seen that the speed with which an increment of the surface enclosing the volume $v(t)$ moves normal to itself is given by $v_i n_i$. Hence the increment of volume Δv swept out in time Δt by this element is given as $v_i n_i\, \Delta t\, da$ and the integral over Δv in equation (3.9) can thus be written as

$$\int_{\Delta v} \left(B + \frac{\partial B}{\partial t}\, \Delta t \right) dv = \oint_a \left(B + \frac{\partial B}{\partial t}\, \Delta t \right) n_i v_i\, \Delta t\, da \tag{3.10}$$

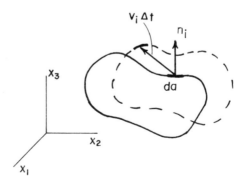

FIGURE 3.3.

On substituting equation (3.10) into (3.9), dividing by Δt, and taking the limit as $\Delta t \to 0$, we thus find

$$\frac{d}{dt} \int_{v(t)} B \, dv = \int_{v(t)} \frac{\partial B}{\partial t} \, dv + \oint_{a(t)} B v_i n_i \, da \tag{3.11}$$

3.5. Green's Theorem

In addition to the above evaluation of the time derivative of a volume integral, it will also be convenient in examining the above mass-conservation and momentum-balance equations to make use of what is called Green's theorem.

If B denotes any scalar or scalar component of a vector or tensor, Green's theorem yields the following relation between surface and volume integrals:

$$\oint_a B n_i \, da = \int_v \frac{\partial B}{\partial x_i} \, dv \tag{3.12}$$

provided B and its first derivatives are continuous.

To show the validity of this relation, let us consider a volume bounded above and below by general curved surfaces and on the sides by cylindrical surfaces having generators parallel to the x_3 axis. Such a volume is shown in Figure 3.4.

In rectangular coordinates, the right-hand side of equation (3.12) is expressible, for $i = 3$, as

$$\int_v \frac{\partial B}{\partial x_3} \, dv = \iiint_v \frac{\partial B}{\partial x_3} \, dx_3 \, dx_1 \, dx_2 \tag{3.13}$$

Integrating from the bottom to the top of the volume, we have

$$\iiint \frac{\partial B}{\partial x_3} \, dx_3 \, dx_1 \, dx_2 = \iint (B^T - B^B) \, dx_1 \, dx_2 \tag{3.14}$$

where B^T and B^B denote the values of B for any x_1 and x_2 at the corresponding top and bottom points of the volume.

Now, for the top surface element having normal n_i^T, as shown in Figure 3.4, we have

$$dx_1 \, dx_2 = n_3^T \, da \tag{3.15}$$

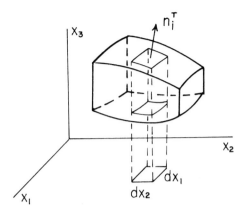

FIGURE 3.4.

and for the bottom surface element having normal n_i^B, we have

$$dx_1 \, dx_2 = -n_3^B \, da \qquad (3.16)$$

the negative sign being due to the downward direction of the normal. On substituting these relations into equation (3.14), we thus find that the integral on the right-hand side of this equation becomes equal to the sum of the integrals of B over the top and bottom surfaces. Moreover, since the sides of the volume are cylindrical surfaces with generators parallel to the x axis, the normal n_3 on these surfaces will be zero, so that equation (3.14) may be written with the help of equations (3.15) and (3.16) as

$$\int_v \frac{\partial B}{\partial x_3} \, dv = \oint_a B n_3 \, da \qquad (3.17)$$

This relation holds for any volumes of the shape shown in Figure 3.4. By addition, it will also hold for any region that can be cut into a finite number of volumes of this kind. Using a limiting process, it follows further that the relation of equation (3.17) applies also for any ordinary region, which can obviously be regarded as the limit of the sum of such parts.

On repeating the above arguments for the x_1 and x_2 integration directions, we may thus establish the validity of equation (3.12).

We note that, by addition, equation (3.12) also applies for vector components B_i such that

$$\oint_a B_i n_i \, da = \int_v \frac{\partial B_i}{\partial x_i} \, dv \qquad (3.18)$$

and for second-order tensor components B_{ji} such that

$$\oint_a B_{ji}n_i \, da = \int_v \frac{\partial B_{ji}}{\partial x_i} \, dv \tag{3.19}$$

3.6. The Stress Vector

We now wish to inquire further into the nature of the stress vector $t_i = t_i(x_j, t, n_j)$ defined by equation (3.3). To do this we first consider a small volume element of material about some point x_i in the continuum. Let the faces of this element be parallel to the coordinate planes and on each of the three positive faces (i.e., faces whose outward normals point in the positive coordinate directions) resolve the associated stress vectors into components along the coordinate directions as shown in Figure 3.5. The notation used in Figure 3.5 in denoting the components of the stress vectors acting on each of the positive faces is as follows: The first subscript denotes the coordinate axis along which the outward normal to the face under consideration points and the second subscript denotes the direction in which the component acts. Thus, for example, t_{12} denotes the x_2 component of the stress vector acting on the faces with normal along x_1.

In order to examine the nature of the stress-vector components acting on the three negative faces of the element, let us apply the principle of linear momentum balance as expressed by equation (3.5). In particular, if l denotes some characteristic length of the element such as Δx_1, Δx_2, Δx_3 or their average value, the volume of the element will clearly be of the order of l^3 while its surface area will be of the order of l^2. As the element size is reduced to zero, its volume will therefore be an order of magnitude smaller than its surface area. On applying to this small element the balance of linear

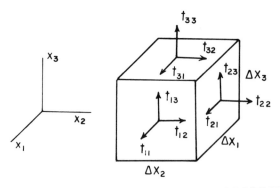

FIGURE 3.5.

momentum as expressed by equations (3.5) and (3.11) and using Green's theorem (assuming continuous velocity components), it is thus seen that the contributions from the volume integrals appearing in this expression will be an order of magnitude smaller than that arising from the surface integral, so that, to first order of smallness of the element, the balance of linear momentum reduces simply to

$$\oint_{a(t)} t_i \, da = 0 \qquad (3.20)$$

This very important result states that the stress vector components acting on the faces of a small element of material yield to a first approximation a system of forces having zero resultant. Of course, as the element is taken smaller and smaller, the approximation simultaneously becomes better and better.

Suppose now that the dimension Δx_1 of the element shown in Figure 3.5 is reduced to zero while holding the dimensions Δx_2 and Δx_3 small but nonzero. In this case, the volume of the element is itself reduced to zero and the only nonzero surface of the remaining element is the x_1 face, having normal along the x_1 direction. Since we may chose this area as small as we please, the only way the above integral, equation (3.20), of the stress vector all the way around this area can vanish is for the stress vector at a point on the negative x_1 face of the element to be equal in magnitude and opposite in direction to the stress vector acting at the same point on the positive x_1 face. This means, therefore, that the components of the stress vector acting at a point on the negative x_1 face must be equal in magnitude and opposite in direction to the components acting on the positive face.

The above argument is, of course, likewise valid for the remaining faces of the element, so that the stress components acting on the negative faces of the element shown in Figure 3.5 are to a first approximation all equal in magnitude and opposite in direction to the components acting on the positive faces.

Since the choice of axes and the associated element orientation are arbitrary, it follows also from the above reasoning that at time t and at point x_i on any real or imagined continuum surface having unit outward normal n_i, the stress vectors acting on opposite sides of this surface are equal in magnitude and opposite in direction. This fact may be expressed mathematically simply as

$$t_i(x_j, t, n_j) = -t_i(x_j, t, -n_j) \qquad (3.21)$$

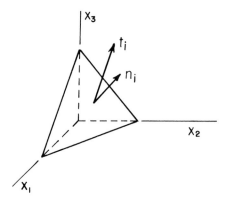

FIGURE 3.6.

3.7. The Stress Tensor

Consider again the element of material shown in Figure 3.5. If we divide the element by passing a diagonal plane through it and consider the remaining part with negative x_1, x_2, and x_3 faces, we have the tetrahedron shown in Figure 3.6. According to equation (3.20), the stress vectors acting on this element form, to a first approximation, a system of forces having zero resultant. If Δa denotes the surface area of the inclined face, the balance of forces in the x_1 direction may thus be written as

$$t_1 \Delta a = t_{11} \Delta a_1 + t_{21} \Delta a_2 + t_{13} \Delta a_3$$

where, from the previous argument, the stress components on the negative coordinate faces, of area Δa_i, are taken equal in magnitude and opposite in direction to those acting on the positive faces. Now, geometry gives

$$\Delta a_i = \Delta a \, n_i$$

where n_i denotes the direction cosines of the outward normal to the inclined face of the tetrahedron. Thus, the above equation becomes

$$t_1 = t_{11} n_1 + t_{21} n_2 + t_{31} n_3$$

Similar equations likewise hold for the remaining two coordinate directions, so that, in the limit as the element shrinks to zero, the equilibrium condition of equation (3.20) at any point can therefore be expressed in tensor notation simply as

$$t_i = t_{ji} n_j \tag{3.22}$$

Thus, the stress-vector components acting at a point on a real or imagined continuum surface are directly related to the components t_{ji} defined in Figure 3.5 through the components of the unit normal.

Since t_i and n_i denote vectors, it follows from the quotient rule of Chapter 1 that the components t_{ij} are components of a second-order Cartesian tensor. This tensor is called the *Cauchy stress tensor* and varies, in general, from point to point and time to time in the continuum; i.e.,

$$t_{ij} = t_{ij}(x_k, t) \tag{3.23}$$

Equations (3.21) and (3.22) represent mathematical statements of the so-called Cauchy stress law. It is important to emphasize that these equations hold for dynamic as well as static loadings even though the equilibrium condition of equation (3.20) was used to obtain them. This, it will be remembered, is due to the fact that inertia and body force contributions to the balance of linear momentum become negligible for small volume elements.

3.8. Change of Stress Components with Rigid Rotations

In connection with the above results, it is worth noticing also that when the balance of angular momentum, as expressed by equation (3.6), is applied to the small element of Figure 3.5, the contribution from the volume integrals in this expression, like those in the linear momentum expression, will be an order of magnitude smaller than the contribution from the surface integral, so that, to first order in smallness of the element, the balance of angular momentum may be written as

$$\oint_a e_{ijk}x_j t_k \, da = 0 \tag{3.24}$$

This result shows that the stress vectors acting on the faces of a small element of material yield, to a first approximation, a system of forces having zero resultant moment. When combined with the earlier result of equation (3.20) requiring that these forces have zero resultant, it is thus seen that these stress forces constitute nothing more than a system of forces in equilibrium.

Consider now the case where the small volume element of Figure 3.5 is subjected to a time-dependent, rigid-body rotation described by the orthogonal tensor

$$Q_{ij} = Q_{ij}(t) \tag{3.25}$$

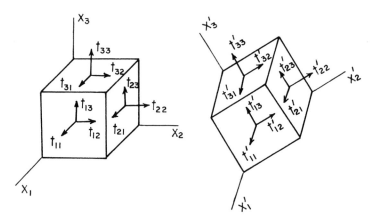

FIGURE 3.7.

having components at time t_0 given by

$$Q_{ij}(t_0) = \delta_{ij} \tag{3.26}$$

The motion of this isolated element is illustrated in Figure 3.7.

Since the stresses acting on this element form, to a first approximation, a system of forces in equilibrium, these stresses, as measured relative to a set of axes x_i' which rotate with the element, will be unaffected by the rotation, and the stress components t_{ij}' at time t will thus be related to the components t_{ij} at time t_0 simply by the coordinate transformation relation for second-order tensors, namely

$$t_{ij}' = Q_{ip}(t)Q_{jk}(t)t_{pk} \tag{3.27}$$

It is worth noting that equation (3.27) does not follow *just* because the stress t_{ij} is a second-order tensor quantity, since tensor definitions, by themselves, do not involve time-dependent transformations.

3.9. Local Form of Mass Conservation

Consider again the principle of mass conservation as expressed by equation (3.2). Using the result of equation (3.11) with this equation, we have

$$\int_v \frac{\partial \varrho}{\partial t}\,dv + \oint_a \varrho v_i n_i\,da = 0 \tag{3.28}$$

Applying Green's theorem to the surface integral appearing in this equation, we find further that the relation can be written as

$$\int_v \left(\frac{\partial \varrho}{\partial t} + \varrho \frac{\partial v_i}{\partial x_i} + v_i \frac{\partial \varrho}{\partial x_i} \right) dv = 0 \tag{3.29}$$

If the above integrand is continuous, the only way this equation can hold for arbitrary choices of the continuum volume v is for the integrand itself to vanish. This condition yields the so-called *continuity equation*

$$\frac{\partial \varrho}{\partial t} + v_i \frac{\partial \varrho}{\partial x_i} + \varrho \frac{\partial v_i}{\partial x_i} = 0 \tag{3.30}$$

expressing the local (differential) form of the principle of conservation of mass.

If we use the material time derivative

$$\frac{d}{dt} = \frac{\partial}{\partial t} + v_j \frac{\partial}{\partial x_j}$$

as given by equation (2.11) of Chapter 2, we may also express this equation in the alternate form

$$\frac{d\varrho}{dt} + \varrho \frac{\partial v_i}{\partial x_i} = 0 \tag{3.31}$$

3.10. Local Form of Linear Momentum Balance

We now consider the balance of the linear momentum as expressed by equation (3.5) in the same fashion as just done for the equation of mass conservation. In particular, on substituting the Cauchy stress law of equation (3.22) into equation (3.5) and evaluating the time derivative in accordance with equation (3.11), we find with the help of Green's theorem that the principle of linear momentum balance of equation (3.5) is expressible as

$$\int_v \left[\frac{\partial}{\partial t} (\varrho v_i) + \frac{\partial}{\partial x_i} (\varrho v_i v_j) - \frac{\partial}{\partial x_j} t_{ji} - f_i \right] dv = 0 \tag{3.32}$$

For a continuous integrand and arbitrary choice of volume, this equation requires

$$\frac{\partial}{\partial t} (\varrho v_i) + \frac{\partial}{\partial x_j} (\varrho v_i v_j) - \frac{\partial t_{ji}}{\partial x_j} - f_i = 0 \tag{3.33}$$

Expanding the derivatives in this equation and using equation (3.30), we can express the local form of linear momentum balance as

$$\varrho \frac{\partial v_i}{\partial t} + \varrho v_j \frac{\partial v_i}{\partial x_j} = \frac{\partial t_{ji}}{\partial x_j} + f_i \tag{3.34}$$

Alternatively, if we again use the material time derivative

$$\frac{d}{dt} = \frac{\partial}{\partial t} + v_j \frac{\partial}{\partial x_j}$$

we may also express this equation simply as

$$\varrho \frac{dv_i}{dt} = \frac{\partial t_{ji}}{\partial x_j} + f_i \tag{3.35}$$

3.11. Local Form of Angular Momentum Balance

If we substitute the Cauchy stress law of equation (3.22) into equation (3.6) expressing the balance of angular momentum and proceed in the same fashion as previously, we find the principle of angular momentum balance expressible as

$$\int_v e_{ijk} \left[\frac{\partial}{\partial t} (x_j \varrho v_k) + \frac{\partial}{\partial x_l} (x_j \varrho v_k v_l) - \frac{\partial}{\partial x_l} (x_j t_{kl}) - x_j f_k \right] dv = 0 \tag{3.36}$$

For a continuous integrand and arbitrary volume, this equation can be satisfied only if the integrand vanishes as in the above previous cases. Moreover, using the linear momentum statement of equation (3.33), the vanishing of this integrand reduces simply to

$$e_{ijk} [\varrho v_j v_k - t_{jk}] = 0 \tag{3.37}$$

By direct expansion it can easily be seen that the first part of this expression is zero, that is

$$e_{ijk} \varrho v_j v_k = 0$$

In these circumstances, equation (3.37) thus reduces to

$$e_{ijk} t_{jk} = 0 \tag{3.38}$$

For $i = 1$, this equation yields immediately that

$$t_{23} - t_{32} = 0$$

Similar equations follow for $i = 2$ and $i = 3$, so that equation (3.38) therefore requires that

$$t_{ij} = t_{ji} \qquad (3.39)$$

The local form of the principle of angular momentum balance thus becomes simply an expression requiring the Cauchy stress tensor to be symmetric.

3.12. Some Simple Examples of Stress

Pressure. When the stress tensor at a point is expressible as

$$t_{ij} = -p\,\delta_{ij} \qquad (3.40)$$

the stress is characterized completely by the scalar p, which is referred to as the *pressure*. In this case, the Cauchy stress law of equation (3.22) yields

$$t_i = t_{ji}n_j = -pn_i \qquad (3.41)$$

thus showing that the stress vector acts normal to every plane passing through the point where equation (3.40) holds.

A small parallelepiped element of material at time t and place x_i is accordingly subjected to equal stress values on all faces as shown in Figure 3.8.

Plane Stress. For the case where the stress tensor components are expressible at a point as

$$
\begin{array}{lll}
t_{11} = t_{11}, & t_{12} = t_{12}, & t_{13} = 0 \\
t_{21} = t_{12}, & t_{22} = t_{22}, & t_{23} = 0 \\
t_{31} = 0, & t_{32} = 0, & t_{33} = 0
\end{array}
\qquad (3.42)
$$

the stress at the point is said to characterize a state of *plane stress* with respect to the plane defined by the x_1 and x_2 axes.

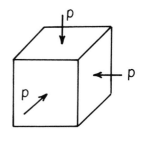

FIGURE 3.8.

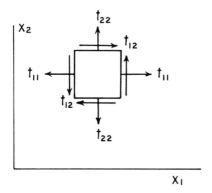

FIGURE 3.9.

Such a state of plane stress is illustrated in Figure 3.9 for a small element of material at time t and place x_i.

It is instructive to examine the change in these stress components with change in orientation of coordinate axes. Introducing such axes x_1', x_2' and taking θ to denote the angle between the x_1 and x_1' axes, as shown in Figure 3.10, we have from Chapter 1 the tensor transformation equation relating the stress components t_{ij}' and t_{ij} expressible as

$$t_{ij}' = a_{im} a_{jn} t_{mn}$$

with the direction cosines a_{ij} given from Figure 3.10 as

$$[a_{ij}] = \begin{bmatrix} \cos\theta & \sin\theta & 0 \\ -\sin\theta & \cos\theta & 0 \\ 0 & 0 & 1 \end{bmatrix}$$

Using equations (3.42), we thus find

$$\begin{aligned} t_{11}' &= t_{11}\cos^2\theta + t_{22}\sin^2\theta + 2t_{12}\sin\theta\cos\theta \\ t_{22}' &= t_{11}\sin^2\theta + t_{22}\cos^2\theta - 2t_{12}\sin\theta\cos\theta \\ t_{12}' &= (t_{22} - t_{11})\sin\theta\cos\theta + t_{12}(\cos^2\theta - \sin^2\theta) \end{aligned} \tag{3.43}$$

These stress components are illustrated in Figure 3.10.

We note that the value of θ defining *principal axes* can be found immediately from the last of equations (3.43) by simply setting t_{12} equal to zero and solving for θ. We find the result expressible as

$$\tan 2\theta = \frac{2t_{12}}{t_{11} - t_{22}} \tag{3.44}$$

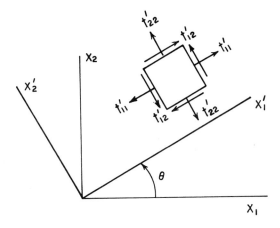

FIGURE 3.10.

On substituting this value of θ into the first two expressions of equation (3.43), we may, of course, also determine the values of the diagonal (principal) stress components t_{11} and t_{22}. After some reduction these are found to be expressible as

$$t'_{11} = \frac{t_{11} + t_{22}}{2} + \frac{1}{2} [(t_{11} - t_{22})^2 + 4t_{12}^2]^{1/2}$$

$$t'_{22} = \frac{t_{11} + t_{22}}{2} - \frac{1}{2} [(t_{11} - t_{22})^2 + 4t_{12}^2]^{1/2}$$

(3.45)

3.13. Stress Boundary Conditions

We now consider the conditions imposed on the stress tensor at some point on the boundary of the continuum where the stress vector is specified. This situation is illustrated in Figure 3.11. With n_i denoting the components

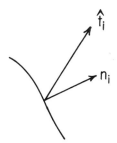

FIGURE 3.11.

of the unit outward normal to the boundary at some point, we may immediately associate the applied stress vector components $\hat{t}_i$ with the components t_i considered earlier in connection with the Cauchy tetrahedron argument; that is,

$$t_i(x_j, t, n_j) = \hat{t}_i(x_j, t, n_j) \tag{3.46}$$

Hence, with equation (3.22) we thus have for a point on the boundary the relation

$$\hat{t}_i = t_{ji}n_j \tag{3.47}$$

The specification of the stress vector at a point on the boundary of the continuum thus imposes through equation (3.47) three conditions on the components of the stress tensor at that point.

Example. Consider the beam of thickness b shown in Figure 3.12 and write the boundary conditions for the upper boundary defined by $x_2 = h$. At every point on the surface $x_2 = h$, the stress vector has components

$$\hat{t}_1 = 0, \qquad \hat{t}_2 = -w_0/b, \qquad \hat{t}_3 = 0$$

The outward normal to this surface likewise has components everywhere given by

$$n_1 = 0, \qquad n_2 = 1, \qquad n_3 = 0$$

Hence, equation (3.47) requires at every point on the boundary $x_2 = h$ that

$$t_{21} = t_{23} = 0$$

and that

$$t_{22} = -w_0/b$$

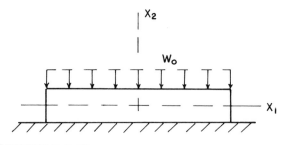

FIGURE 3.12.

3.14. Approximations Valid for Small Deformations

For small deformations, we may assume, in accordance with the discussion of Chapter 2, that the displacement gradients are all small in comparison with unity; that is, that

$$\frac{\partial u_i}{\partial X_j} \ll 1 \tag{3.48}$$

Under this restriction, we also have from Chapter 2 that derivatives with respect to x_i are indistinguishable from derivatives with respect to X_i. In this case, we may, accordingly, express the local form of mass conservation, as given by equation (3.31), as

$$\frac{d\varrho}{dt} + \varrho \frac{\partial v_i}{\partial X_i} = 0 \tag{3.49}$$

Noticing further from the results of Chapter 2 that

$$\frac{d\varrho}{dt} = \frac{\partial \varrho}{\partial t} (X_j, t) \tag{3.50}$$

it is seen that the first term of equation (3.49) represents nothing more than the time derivative of the density, holding the material coordinates X_j fixed. Also, the last term in this equation can be expressed in terms of the displacement components u_i as

$$\frac{\partial v_i}{\partial X_i} = \frac{\partial^2 u_i}{\partial X_i \partial t} = \frac{\partial^2 u_i}{\partial t \partial X_i} \tag{3.51}$$

Under these circumstances, equation (3.49) thus becomes expressible as

$$\frac{\partial \varrho}{\partial t} + \varrho \frac{\partial^2 u_i}{\partial t \partial X_i} = 0 \tag{3.52}$$

which, on integrating, gives

$$\frac{\varrho}{\varrho_0} = \exp\left(-\frac{\partial u_i}{\partial X_i}\right) \tag{3.53}$$

Finally, expressing the density ϱ in terms of the initial density ϱ_0 and a change in density ϱ' as

$$\varrho = \varrho_0 + \varrho' \tag{3.54}$$

and expanding the exponential terms in equation (3.53) in a Taylor series, we thus have, for small deformations, that

$$\frac{\varrho'}{\varrho_0} = -\frac{\partial u_i}{\partial X_i} \tag{3.55}$$

On treating, where necessary, the balance of linear momentum given by equation (3.35) in an analogous fashion, we may thus finally write the equations for mass conservation and momentum balance for small deformations as

$$\varrho' = -\varrho_0 \frac{\partial u_i}{\partial X_i} \tag{3.56}$$

$$\varrho_0 \frac{\partial^2 u_i}{\partial t^2} = \frac{\partial t_{ji}}{\partial X_j} + f_i \tag{3.57}$$

$$t_{ij} = t_{ji} \tag{3.58}$$

3.15. Unabridged Notation

If we adopt the unabridged notation introduced in Chapter 2, we have for initial coordinates

$$x = X_1, \qquad y = X_2, \qquad z = X_3 \tag{3.59}$$

and for displacement components

$$u = u_1, \qquad v = u_2, \qquad w = u_3 \tag{3.60}$$

In addition, we denote the stress components as

$$t_{xx} = t_{11}, \qquad t_{xy} = t_{12}, \qquad \text{etc.} \tag{3.61}$$

and the body force components as

$$f_x = f_1, \qquad f_y = f_2, \qquad f_z = f_3 \tag{3.62}$$

Using this notation, the above equation for mass conservation under the restriction of small deformations is expressible as

$$\varrho' = -\varrho_0 \left(\frac{\partial u}{\partial x} + \frac{\partial v}{\partial y} + \frac{\partial w}{\partial z} \right) \tag{3.63}$$

The equations for linear momentum balance for small deformations are likewise expressible as

$$\varrho_0 \frac{\partial^2 u}{\partial t^2} = \frac{\partial t_{xx}}{\partial x} + \frac{\partial t_{yx}}{\partial y} + \frac{\partial t_{zx}}{\partial z} + f_x \tag{3.64}$$

$$\varrho_0 \frac{\partial^2 v}{\partial t^2} = \frac{\partial t_{xy}}{\partial x} + \frac{\partial t_{yy}}{\partial y} + \frac{\partial t_{zy}}{\partial z} + f_y \tag{3.65}$$

$$\varrho_0 \frac{\partial^2 w}{\partial t^2} = \frac{\partial t_{xz}}{\partial x} + \frac{\partial t_{yz}}{\partial y} + \frac{\partial t_{zz}}{\partial z} + f_z \tag{3.66}$$

and the equations for angular momentum balance are expressible as

$$t_{xy} = t_{yx}, \qquad t_{xz} = t_{zx}, \qquad t_{yz} = t_{zy} \tag{3.67}$$

3.16. Cylindrical Polar Coordinates

Using the same general procedure as introduced in Chapter 2, we may transform the above equations expressing mass conservation and momentum balance to cylindrical polar coordinates r, θ, z.

In order to define the stress components at a point P in polar coordinates, we consider, as in Chapter 2, a new set of rectangular axes x', y', z' with z' along z and x' and y' along the directions of increasing r and θ (Figure 3.13). As before, the polar components of displacement are given by

$$u_r = u', \qquad u_\theta = v', \qquad u_z = w' \tag{3.68}$$

where u', v', and w' denote the rectangular displacement components relative to the x', y', z' axes.

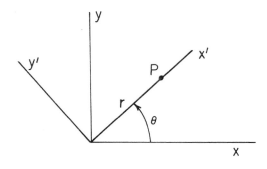

FIGURE 3.13.

Similarly, the Cauchy stress components in polar coordinates are defined as

$$t_{rr} = t'_{xx}, \qquad t_{r\theta} = t'_{xy}, \qquad t_{rz} = t'_{xz}$$
$$t_{\theta r} = t'_{yx}, \qquad t_{\theta\theta} = t'_{yy}, \qquad t_{\theta z} = t'_{yz} \qquad (3.69)$$
$$t_{zr} = t'_{zx}, \qquad t_{z\theta} = t'_{zy}, \qquad t_{zz} = t'_{zz}$$

and the body force components as

$$f_r = f'_x, \qquad f_\theta = f'_y, \qquad f_z = f'_z \qquad (3.70)$$

In terms of these components, the equations expressing mass conservation and momentum balance for small deformations are then found expressible by the methods of Chapter 2 as

$$\varrho' = \varrho_0\left(\frac{\partial u_r}{\partial r} + \frac{1}{r}\frac{\partial u_\theta}{\partial \theta} + \frac{u_r}{r} + \frac{\partial u_z}{\partial z}\right) \qquad (3.71)$$

$$\varrho_0\frac{\partial^2 u_r}{\partial t^2} = \frac{\partial t_{rr}}{\partial r} + \frac{1}{r}\frac{\partial t_{\theta r}}{\partial \theta} + \frac{t_{rr} - t_{\theta\theta}}{r} + \frac{\partial t_{zr}}{\partial z} + f_r \qquad (3.72)$$

$$\varrho_0\frac{\partial^2 u_\theta}{\partial t^2} = \frac{\partial t_{r\theta}}{\partial r} + \frac{1}{r}\frac{\partial t_{\theta\theta}}{\partial \theta} + \frac{2t_{r\theta}}{r} + \frac{\partial t_{z\theta}}{\partial z} + f_\theta \qquad (3.73)$$

$$\varrho_0\frac{\partial^2 u_z}{\partial t^2} = \frac{\partial t_{rz}}{\partial r} + \frac{1}{r}\frac{\partial t_{\theta z}}{\partial \theta} + \frac{t_{rz}}{r} + \frac{\partial t_{zz}}{\partial z} + f_z \qquad (3.74)$$

$$t_{r\theta} = t_{\theta r}, \qquad t_{rz} = t_{zr}, \qquad t_{\theta z} = t_{z\theta} \qquad (3.75)$$

Selected Reading

Malvern, L. E., *Introduction to the Mechanics of a Continuous Medium*. Prentice-Hall, Englewood Cliffs, New Jersey, 1969. Chapters 3 and 5 give detailed discussions of stress and governing continuum equations. Section 5.3 also provides a discussion of the governing equations referred to material coordinates rather than to spatial coordinates, as employed here.

Frederick, D., and T. S. Chang, *Continuum Mechanics*. Allyn and Bacon, Boston, Massachusetts, 1965. Chapter 2 provides a discussion of stress and Chapter 4 gives a concise derivation of the local forms of mass conservation and momentum balance.

Fung, Y. C., *Foundations of Solid Mechanics*. Prentice-Hall, Englewood Cliffs, New Jersey, 1965. Chapters 3 and 5 give thorough discussions of stress and conservation and balance laws.

Long, R. R., *Mechanics of Solids and Fluids*. Prentice-Hall, Englewood Cliffs, New Jersey, 1961. Chapter 2 introduces the concept of stress and provides an elementary derivation of the local forms of the balance of linear and angular momentum.

Truesdell, C., and R. A. Toupin, "The Classical Field Theories," in *Encyclopedia of Physics*, Vol III/1. Springer-Verlag, Berlin, 1960. Chapter B discusses conservation of mass and Chapter D presents the general theory of momentum.

Truesdell, C., *Essays in the History of Mechanics*. Springer-Verlag, Berlin, 1968. Beautifully written essays on the early history of mechanics and the development of its general principles.

Exercises

1. Let t_{ij}^A and t_{ij}^B denote stress components in two bodies A and B which are welded together along an interface having unit normal vector pointing toward body B with components $n_i(x_j, t)$. Show that at any point on this interface, the stress components must satisfy

    $$(t_{ji}^A - t_{ji}^B)n_j = 0$$

2. A shock wave is defined as a surface in a continuum across which one or more components of the stress and velocity are discontinuous. Using equation (3.11) with (3.2) and (3.5), show (by arguments similar to those employed in Section 3.6) that for a *stationary shock*, the following relations must apply at every point on the shock surface:

    $$[\varrho v_i]n_i = 0, \qquad [\varrho v_i v_j - t_{ji}]n_j = 0$$

    where brackets are used to denote the difference in the enclosed quantity immediately ahead of and behind the shock surface, and n_i denotes components of the normal to the surface. Note that, for this case, Green's theorem cannot be applied to the surface integral in equation (3.11).

3. Show in Exercise 2 that no additional restrictions arise from consideration of the balance of angular momentum.

4. By substituting $v_k - Un_k$ for v_k, show that the results of Exercise 2 can be generalized to include the case where the shock surface is moving with constant speed U along its normal.

5. Assuming equilibrium and the absence of body forces, show that the average stress $\bar{t}_{ij}$ over a region v with a surface a is expressible as

    $$\bar{t}_{ij} = \frac{1}{v} \oint_a t_j x_i \, da$$

    where t_j denotes components of the Cauchy stress vector and x_i denotes position in the region v.

6. Show, using the plane-stress equations of Section 3.12, that the normal stress components t_{11} and t_{22} have stationary values (maxima or minima) for principal axes.

7. Determine, using the plane-stress equations of Section 3.12, the angle θ between the x_1 and x_1' axes for which the shear stress t_{12} has stationary values. Express these values in terms of t_{11}, t_{12}, and t_{22}.

PART *II*

CLASSICAL ELASTICITY

The theory of elasticity is concerned with the mechanics of bodies made of elastic, springy materials which deform on loading and recover completely on unloading. By combining assumptions regarding the stress and deformation of an elastic material with the governing continuum equations of momentum balance, the theory attempts to reduce to calculation all questions concerning the internal stresses and motions of an elastic body subjected to general loading conditions.

The theory had its beginning in England with *Robert Hooke*, who in 1676 proposed on the basis of experiment that the force and extension of an elastic body were proportional to one another. The theory was further developed on the European continent by *James Bernoulli* and *Leonhard Euler* during the eighteenth century in their mathematical studies of the bending of beams. The extension of the mathematical theory to more general solids was first made by *Louis Navier* in 1821 using special assumptions concerning the molecular forces of elastic solids. The following year *Augustin Cauchy*, without regard to molecular considerations, introduced the concepts of stress and small deformation at a point and assumed a linear relation between them, to give the general theory of linear elasticity. Technical applications began in earnest in 1855, when the mathematician *Barre de Saint-Venant* solved the problem of the twisting of prismatic bars and worked out detailed numerical results.

4

Theory of Elasticity

An inspection of the mechanical laws described in the last chapter reveals one scalar equation expressing the conservation of mass, three scalar equations expressing the balance of linear momentum, and three scalar equations expressing the balance of angular momentum, thus giving a total of seven equations in all. If we count the number of unknowns introduced into these equations, we find, however, one scalar density function, three scalar components of displacement or velocity, and nine scalar components of stress, so that, assuming the body forces are given, we thus have 13 unknowns and only seven equations connecting them. To determine the unknowns, it is therefore necessary to employ six additional equations. These equations provide information on the response of the particular material under consideration and are known as *constitutive relations*.

In the present chapter, we consider such constitutive relations for a so-called *elastic solid*. After developing these relations in their general non-linear form, we restrict attention to the case of small deformations and combine the resulting linearized constitutive relations with the equations of momentum balance to obtain governing equations associated with the classical *linear theory of elasticity*.

4.1. Constitutive Relations for an Elastic Solid

For an elastic solid, the constitutive relations are assumed to connect the stress acting on an element of material at time t with its deformation from an unstressed *natural* or *reference configuration*. With respect to Figure 4.1, we may thus write the constitutive relation for an elastic solid in terms

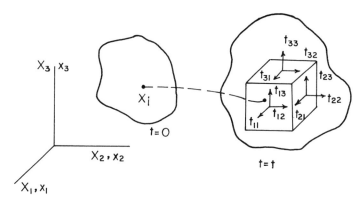

FIGURE 4.1.

of the Cauchy stress components t_{ij} as

$$t_{ij} = \phi_{ij}\left(\frac{\partial x_k}{\partial X_l}\right) \tag{4.1}$$

where ϕ_{ij} denotes a tensor function which is symmetric in accordance with the angular momentum requirements and where the material coordinates X_l refer to the reference configuration of the solid such that when the deformation is zero, so is the stress.

In general, the constitutive relation of equation (4.1) may, of course, vary from one material particle to another; that is, the function ϕ_{ij} may depend also on the material coordinates X_i. In this case, we speak of the material as an *inhomogeneous material*. Alternatively, when the constitutive relation is of the same form for all particles in the body, we then speak of the material as a *homogeneous material*.

It is interesting to note that equation (4.1) expresses a three-dimensional continuum version of the well-known elastic concept associated with springs, namely that the application of a force gives rise to a deflection which subsequently vanishes again on removal of the loading. In this case, the reference configuration of the elastic body is clearly the unstretched, free length of the spring.

4.2. Restrictions Placed on Constitutive Relations by Principle of Material Indifference

It is assumed that the basic form of the constitutive relation of equation (4.1) will remain unchanged under arbitrary time-dependent, rigid-body translations and rotations of the element of material shown in Figure 4.1 such that if

$$t_{ij} = \phi_{ij}\left(\frac{\partial x_k}{\partial X_n}\right) \tag{4.2}$$

is to hold for the motion

$$x_k = x_k(X_n, t) \tag{4.3}$$

then the relation

$$t'_{ij} = \phi_{ij}\left(\frac{\partial x'_k}{\partial X_n}\right) \tag{4.4}$$

must also hold for the motion

$$x'_k = Q_{kj}x_j + U_k \tag{4.5}$$

where U_k and Q_{kj} describe the arbitrary time-dependent, rigid-body translation and rotation of the solid and where t'_{ij} and $\partial x'_k/\partial X_n$ are given from equation (3.27) of Chapter 3 and equation (4.5) as

$$t'_{ij} = Q_{ik}Q_{jl}t_{kl} \tag{4.6}$$

and

$$\frac{\partial x'_k}{\partial X_n} = Q_{km}\frac{\partial x_m}{\partial X_n} \tag{4.7}$$

If we combine equations (4.4), (4.6), and (4.7), we can easily see that the above condition requires that the constitutive relation of equation (4.1) satisfy

$$Q_{ik}Q_{jl}t_{kl} = \phi_{ij}\left(Q_{km}\frac{\partial x_m}{\partial X_n}\right) \tag{4.8}$$

for all rigid rotations Q_{ij}. Multiplying this equation through by $Q_{ir}Q_{js}$ and using the orthogonality condition appropriate to Q_{ij}, we thus find this relation expressible as

$$t_{rs} = Q_{ir}Q_{js}\phi_{ij}\left(Q_{km}\frac{\partial x_m}{\partial X_n}\right)$$

or, on interchanging the indices i and r and j and s, as

$$t_{ij} = Q_{ri}Q_{sj}\phi_{rs}\left(Q_{km}\frac{\partial x_m}{\partial X_n}\right) \tag{4.9}$$

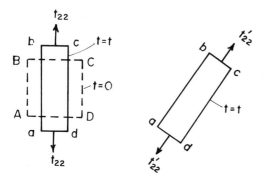

FIGURE 4.2.

This equation clearly imposes a restriction on the form of the elastic constitutive relation of equation (4.1). It expresses, in particular, the *principle of material indifference* to arbitrary time-dependent, rigid-body motions of the material at time t. The concept is illustrated in Figure 4.2, where the stress-deformation response of the element shown is assumed independent of an arbitrary rigid translation and rotation of the element.

To examine the restriction of equation (4.9) in detail, we first recall from equation (2.31) of Chapter 2 that the deformation gradient $F_{ij} = \partial x_i / \partial X_j$ associated with the motion

$$x_i = x_i(X_j, t) \tag{4.10}$$

can be decomposed into a rotation tensor R_{mj} and a stretch tensor U_{jn} such that

$$F_{mn} = R_{mj} U_{jn} \tag{4.11}$$

Since equation (4.9) is to hold for all rotations, it certainly must hold for a rotation equal and opposite to that of the motion of equation (4.10), namely

$$Q_{ij} = R_{ji} \tag{4.12}$$

Thus, from equation (4.9), we have

$$t_{ij} = R_{ir} R_{js} \phi_{rs}(R_{mk} F_{mn}) \tag{4.13}$$

But from equation (4.11) and the fact that R_{ij} is orthogonal, we have also that

$$R_{mk} F_{mn} = U_{kn} \tag{4.14}$$

Hence, equation (4.13) becomes

$$t_{ij} = R_{ir}R_{js}\phi_{rs}(U_{kn}) \tag{4.15}$$

Conversely, since additional rigid-body motions merely change the rotation components R_{ij}, this equation also satisfies equation (4.9) for any rigid rotation. Equation (4.15) therefore provides the form which the constitutive relation of equation (4.1) must take when account is taken of the principle of material indifference.

For an alternate, but equally valid form of the equation, we may recall from Chapter 2 that the stretch U_{kn} is a function only of the deformation tensor C_{pq} defined by

$$C_{pq} = F_{ip}F_{iq} = \frac{\partial x_i}{\partial X_p} \frac{\partial x_i}{\partial X_q} \tag{4.16}$$

so that, on taking a new function $\psi_{mn}(C_{pq})$ such that

$$\phi_{rs} = U_{rm}U_{sn}\psi_{mn}(C_{pq})$$

we can express equation (4.15) with the help of equation (4.11) as

$$t_{ij} = F_{im}F_{jn}\psi_{mn}(C_{pq})$$

or

$$t_{ij} = \frac{\partial x_i}{\partial X_m} \frac{\partial x_j}{\partial X_n} \psi_{mn}(C_{pq}) \tag{4.17}$$

Thus, the condition imposed by equation (4.9) requires that the elastic constitutive relations of equation (4.1) take the restrictive form of equation (4.15) or an equivalent form such as given by equation (4.17). It is worth noticing that, with $\psi_{mn} = \psi_{nm}$ in equation (4.17), we have $t_{ij} = t_{ji}$ as required by angular momentum considerations.

4.3. Material Symmetry Restrictions on the Constitutive Relations

Most elastic materials possess some symmetry in their undeformed state such that their mechanical and physical properties at a point are identical in certain directions. If the properties are identical for all directions from the point, the material is said to be *isotropic*; otherwise it is said to be *anisotropic*.

For example, if we imagine taking a small element of material out of the continuum and testing its properties, we may expect for an isotropic material that these properties will be identical to those which would have been found from the testing of a differently oriented element taken about the same point. If the material is anisotropic with only certain special symmetries, we may, of course, then expect identical properties only when the two differently oriented elements of material are chosen in accord with the material symmetry.

To describe such material symmetries mathematically, we may thus state the existence of a number of rotations through which an element of material can initially be turned during an imagined testing procedure without subsequently detecting any change in its properties. In particular, suppose Q_{ij} denotes the components describing any such allowable symmetry rotation. If we require that

$$t_{ij} = \frac{\partial x_i}{\partial X_m} \frac{\partial x_j}{\partial X_n} \psi_{mn}(C_{pq}) \tag{4.18}$$

is to hold for the motion

$$x_k = x_k(X_j, t) \tag{4.19}$$

then we must also require that

$$t_{ij} = \frac{\partial x_i}{\partial X'_m} \frac{\partial x_j}{\partial X'_n} \psi_{mn}(C'_{pq}) \tag{4.20}$$

hold for

$$x_i = x_i(X'_j, t) \tag{4.21}$$

where

$$X'_j = Q_{jk}X_k \tag{4.22}$$

The above requirements are illustrated in Figure 4.3 for an element of material having a material symmetry represented by a clockwise rotation of $90°$. In this figure, the stress-deformation response at time t (solid lines) will be identical regardless of which initial orientation (dashed lines) is considered.

An alternate but equivalent method of describing the material symmetries is to imagine that, in addition to turning the initial element through the allowed symmetry rotation, an equal rigid-body rotation is also applied to the element at time t. By the principle of material indifference, this latter

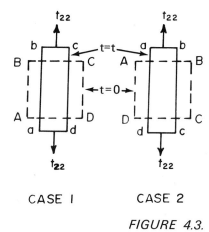

CASE I   CASE 2

FIGURE 4.3.

rotation will, of course, leave the material properties unchanged. Thus, suppose Q_{ij} again denotes the components of any rotation describing the symmetry of the material. If

$$t_{ij} = \frac{\partial x_i}{\partial X_m} \frac{\partial x_j}{\partial X_n} \psi_{mn}(C_{pq}) \tag{4.23}$$

is to hold for the motion

$$x_i = x_i(X_j, t) \tag{4.24}$$

then we also require that

$$t'_{ij} = \frac{\partial x'_i}{\partial X'_m} \frac{\partial x'_j}{\partial X'_n} \psi_{mn}(C'_{pq}) \tag{4.25}$$

hold for

$$x'_i = x_i(X'_j, t) \tag{4.26}$$

where

$$x'_i = Q_{ij}x_j, \qquad X'_j = Q_{jk}X_k, \qquad t'_{ij} = Q_{ik}Q_{jl}t_{kl}, \qquad C'_{pq} = Q_{pk}Q_{ql}C_{kl}$$

The above scheme of representing material symmetries is illustrated in Figure 4.4 for the case of a 90° clockwise allowable symmetry rotation (dashed lines). In this figure, the stress deformation response at time t (solid lines) will be identical for the two cases shown. Notice that case 2 of this figure differs from case 2 of the previous figure by a rigid 90° clockwise rotation of the element and its stress system.

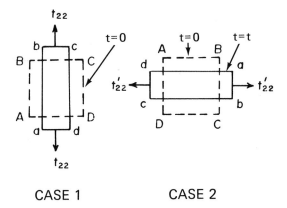

CASE 1   CASE 2

FIGURE 4.4.

4.4. Elastic Constitutive Relations Applicable to Small Deformations

We now wish to examine the form which the elastic constitutive relations of equation (4.17) take when the deformation is assumed to be small. As in Chapter 2, we introduce the displacement u_i such that

$$x_i = X_i + u_i \tag{4.27}$$

and require for small deformation that

$$\frac{\partial u_i}{\partial X_j} \ll 1 \tag{4.28}$$

With the strain tensor components e_{ij} defined in Chapter 2 as

$$e_{ij} = \frac{1}{2}\left(\frac{\partial u_i}{\partial X_j} + \frac{\partial u_j}{\partial X_i}\right) \tag{4.29}$$

we have from equations (2.14) and (2.26) of that chapter that

$$C_{ij} = \delta_{ij} + 2e_{ij} \tag{4.30}$$

so that, on substituting equations (4.27) and (4.30) into equation (4.17), we thus find, under the restriction of small deformations, the following approximate relation:

$$t_{ij} = \psi_{ij}(e_{pq}) \tag{4.31}$$

Moreover, since the displacement gradients are assumed small, the strain components will likewise be small and equation (4.31) may thus be expanded into a Taylor series about $e_{ij} = 0$ to obtain the first-order relation

$$t_{ij} = \psi_{ij}(0) + \left(\frac{\partial \psi_{ij}}{\partial e_{kl}}\right)_0 e_{kl} \tag{4.32}$$

Assuming the stress vanishes when the strain vanishes, this equation may thus finally be written as

$$t_{ij} = C_{ijkl} e_{kl} \tag{4.33}$$

where

$$C_{ijkl} = \left(\frac{\partial \psi_{ij}}{\partial e_{kl}}\right)_0$$

Equation (4.33) denotes the general constitutive relations applicable for small deformations of an elastic solid. The constants C_{ijkl} are referred to as the *elastic constants* of the material. It is easy to show by a direct extension of the quotient rule of Chapter 1 that these constants form the components of a fourth-order Cartesian tensor. In general, there exist, of course, $3^4 = 81$ such constants, but due to the symmetric nature of t_{ij} and e_{ij}, it is easily seen that

$$C_{ijkl} = C_{jikl} = C_{ijlk} = C_{jilk} \tag{4.34}$$

so that only 36 of these constants are actually independent.

4.5. Restriction on Elastic Constants Due to Existence of a Strain Energy Function

In addition to the assumptions leading to the elastic constitutive relation of equation (4.33), it is customary also to assume that the stress is derivable from a scalar function of the strain according to the relation

$$t_{ij} = \frac{\partial U}{\partial e_{ij}} \tag{4.35}$$

where the function

$$U = U(e_{ij}) \tag{4.36}$$

is referred to as the *strain-energy function*.

Expanding this function in a Taylor series about $e_{ij} = 0$ and retaining only second-order terms, we have

$$U = U_0 + \alpha_{ij}e_{ij} + \alpha_{ijkl}e_{ij}e_{kl} \tag{4.37}$$

where U_0, α_{ij}, and α_{ijkl} denote constants. Using equation (4.35), we find the stress given as

$$t_{ij} = \alpha_{ij} + (\alpha_{ijkl} + \alpha_{klij})e_{kl} \tag{4.38}$$

Finally, choosing $\alpha_{ij} = 0$ corresponding to the case of zero initial stress and taking C_{ijkl} as

$$C_{ijkl} = \alpha_{ijkl} + \alpha_{klij} \tag{4.39}$$

we thus have from equation (4.37) that the strain energy function is expressible for $U_0 = 0$ as

$$U = \tfrac{1}{2}C_{ijkl}e_{ij}e_{kl} \tag{4.40}$$

and that the stress is expressible as

$$t_{ij} = C_{ijkl}e_{kl} \tag{4.41}$$

This last equation is, of course, identical in form to that given by equation (4.33). There exists, however, one important difference in the development of the two, namely equation (4.39), which requires that

$$C_{ijkl} = C_{klij} \tag{4.42}$$

Under this restriction, the 36 distinct elastic constants of equation (4.33) can now easily be seen to be reduced to only 21 such constants.

It is worth noting in connection with the above discussion that the assumption leading to equation (4.33) defines what is sometimes called *Cauchy elasticity*, while the additional assumption of equation (4.35) defines what is sometimes referred to as *Green elasticity*.

4.6. Restrictions on Elastic Constants Due to Material Symmetries

We now show that the above elastic constants C_{ijkl} can be further reduced by the presence of material symmetries. In particular, if equation (4.41) is to hold for one orientation of the material, then from equation

(4.25) we also require that it hold for

$$t'_{ij} = C_{ijkl}e'_{kl} \tag{4.43}$$

where

$$t'_{ij} = Q_{im}Q_{jn}t_{mn}, \qquad e'_{kl} = Q_{km}Q_{ln}e_{mn}$$

and where Q_{ij} denotes the components of an allowable symmetry rotation. On substituting the expressions for t'_{ij} and e'_{ij} into equation (4.43) and using the orthogonality conditions appropriate to Q_{ij}, we thus find on comparing with equation (4.41) that C_{ijkl} must satisfy

$$C_{pqmn} = Q_{ip}Q_{jq}Q_{km}Q_{ln}C_{ijkl} \tag{4.44}$$

Example. As an example of the use of the above equation, we consider the case of a material possessing three mutually perpendicular planes of symmetry. Such a material is referred to as an orthotropic material. For this material, we take coordinates axes perpendicular to the three planes of symmetry and require the elastic properties to be unchanged under 180° rotations about these axes.

Consider first the rotation of 180° about the X_3 axis. The components describing this rotation are given by

$$[Q_{ij}] = \begin{bmatrix} -1 & 0 & 0 \\ 0 & -1 & 0 \\ 0 & 0 & 1 \end{bmatrix}$$

Using equation (4.44), we thus find with this rotation the following nontrivial results:

$$C_{1311} = -C_{1311}, \qquad C_{1322} = -C_{1322}$$
$$C_{1333} = -C_{1333}, \qquad C_{1313} = -C_{1313}$$

so that

$$C_{1311} = C_{1322} = C_{1333} = C_{1313} = 0$$

In addition, we also find with the above rotation that

$$C_{2311} = C_{2322} = C_{2333} = C_{2312} = 0$$
$$C_{1213} = C_{1223} = C_{1123} = C_{1113} = 0$$

and

$$C_{2223} = C_{2213} = C_{3323} = C_{3313} = 0$$

In a like manner, we find on considering similar rotations about the X_2 or X_3 axis that

$$C_{1112} = C_{2212} = C_{3312} = 0$$

and

$$C_{2331} = C_{3123} = 0$$

Collecting the above results, we thus have the constitutive relations for an orthotropic material expressible as

$$t_{11} = C_{1111}e_{11} + C_{1122}e_{22} + C_{1133}e_{33}$$
$$t_{22} = C_{2211}e_{11} + C_{2222}e_{22} + C_{2233}e_{33}$$
$$t_{33} = C_{3311}e_{11} + C_{3322}e_{22} + C_{3333}e_{33}$$
$$t_{12} = C_{1212}e_{12}$$
$$t_{23} = C_{2323}e_{23}$$
$$t_{13} = C_{1313}e_{13}$$

It will be seen that the material symmetries have accordingly reduced the 36 original elastic constants of equation (4.33) to 12 independent ones. In addition, from the strain-energy symmetries of equation (4.42), we also have

$$C_{2211} = C_{1122}, \qquad C_{3311} = C_{1133}, \qquad C_{3322} = C_{2233}$$

so that the above 12 elastic constants are reduced further to only nine independent ones.

4.7. Constitutive Relations for Isotropic Elastic Materials

For an *isotropic material*, all possible symmetry rotations Q_{ij} must leave the elastic constants unchanged. Consider, in particular, a 90° rotation about the X_1 axis as defined by

$$[Q_{ij}] = \begin{bmatrix} 1 & 0 & 0 \\ 0 & 0 & 1 \\ 0 & -1 & 0 \end{bmatrix} \tag{4.45}$$

and a 90° rotation about the X_3 axis as defined by

$$[Q_{ij}] = \begin{bmatrix} 0 & 1 & 0 \\ -1 & 0 & 0 \\ 0 & 0 & 1 \end{bmatrix} \tag{4.46}$$

Using each of these allowable symmetry rotations in equation (4.44), we find, after reductions analogous to those considered in the above example, that the constitutive relations reduce to

$$
\begin{aligned}
t_{11} &= C_{1111}e_{11} + C_{1122}e_{22} + C_{1122}e_{33} \\
t_{22} &= C_{1122}e_{11} + C_{1111}e_{22} + C_{1122}e_{33} \\
t_{33} &= C_{1122}e_{11} + C_{1122}e_{22} + C_{1111}e_{33} \\
t_{12} &= C_{1212}e_{12} \\
t_{23} &= C_{1212}e_{23} \\
t_{13} &= C_{1212}e_{13}
\end{aligned}
\tag{4.47}
$$

In addition to the above symmetry rotations, consider also a 45° rotation about the X_3 axis as described by

$$
[Q_{ij}] = \begin{bmatrix} 1/\sqrt{2} & 1/\sqrt{2} & 0 \\ -1/\sqrt{2} & 1/\sqrt{2} & 0 \\ 0 & 0 & 1 \end{bmatrix}
\tag{4.48}
$$

With this additional symmetry restriction, we find further that equation (4.44) requires that

$$
C_{1212} = C_{1111} - C_{1122}
\tag{4.49}
$$

so that the three elastic constants in the above constitutive relations are reduced to only two independent ones. Introducing the notation

$$
\lambda = C_{1122}, \qquad 2\mu = C_{1111} - C_{1122}
\tag{4.50}
$$

these relations are thus seen to be expressible as

$$
\begin{aligned}
t_{11} &= \lambda(e_{11} + e_{22} + e_{33}) + 2\mu e_{11} \\
t_{22} &= \lambda(e_{11} + e_{22} + e_{33}) + 2\mu e_{22} \\
t_{33} &= \lambda(e_{11} + e_{22} + e_{33}) + 2\mu e_{33} \\
t_{12} &= 2\mu e_{12} \\
t_{23} &= 2\mu e_{23} \\
t_{13} &= 2\mu e_{13}
\end{aligned}
\tag{4.51}
$$

In algebraic form, the elastic constants C_{ijkl} of equation (4.33) are easily seen from the above results to be expressible for an isotropic elastic material as

$$
C_{ijkl} = \lambda\, \delta_{ij}\, \delta_{kl} + 2\mu\, \delta_{ik}\, \delta_{jl}
\tag{4.52}
$$

We now show that no further reduction in the number of these constants is possible by symmetry arguments. In particular, if we substitute equation (4.52) into (4.44) and use the definition of δ_{ij}, we find, for any symmetry rotation Q_{ij}, the following expression:

$$C_{ijkl} = \lambda Q_{in} Q_{jn} Q_{kq} Q_{lq} + 2\mu Q_{ip} Q_{kp} Q_{jq} Q_{lq}$$

Using the orthogonality conditions appropriate to any rotation, namely $Q_{in} Q_{jn} = \delta_{ij}$, this relation is easily seen to reduce merely to equation (4.52). This equation is thus invariant under further symmetry rotations and, accordingly, gives the elastic constants appropriate to an isotropic material.

From equation (4.51), we may thus write the constitutive relation for an isotropic elastic material in the following compact form:

$$t_{ij} = \lambda e_{kk} \delta_{ij} + 2\mu e_{ij} \tag{4.53}$$

where $e_{kk} = e_{11} + e_{22} + e_{33}$. The elastic constants λ and μ are referred to as the *Lamé constants*.

4.8. Alternate Form of Elastic Constitutive Relations

We consider the simple tension loading of a cylindrical rod as shown in Figure 4.5. Assuming only $t_{33} \neq 0$, we have from equation (4.53) the following equations:

$$0 = (\lambda + 2\mu)e_{11} + \lambda(e_{22} + e_{33})$$
$$0 = (\lambda + 2\mu)e_{22} + \lambda(e_{11} + e_{33})$$
$$t_{33} = (\lambda + 2\mu)e_{33} + \lambda(e_{11} + e_{22}) \tag{4.54}$$
$$e_{12} = e_{23} = e_{13} = 0$$

t_{33}

t_{33}

FIGURE 4.5.

From the first two of these equations we can easily find that

$$e_{11} = e_{22} = -\nu e_{33} \qquad (4.55)$$

where ν is given by

$$\nu = \frac{\lambda}{2(\lambda + \mu)} \qquad (4.56)$$

and is referred to as *Poisson's ratio*.

Using this result and the third of equations (4.54), we also find

$$t_{33} = E e_{33} \qquad (4.57)$$

where E is given by

$$E = \frac{\mu(3\lambda + 2\mu)}{\lambda + \mu} \qquad (4.58)$$

and is referred to as *Young's modulus*.

By solving for λ and μ from equations (4.56) and (4.58), the elastic constitutive relation for an isotropic material may alternatively be expressed in terms of Young's modulus and Poisson's ratio in the convenient form

$$e_{ij} = \frac{1 + \nu}{E} t_{ij} - \frac{\nu}{E} t_{kk} \delta_{ij} \qquad (4.59)$$

Table 4.1 gives typical values of the above elastic constants for some common materials.

TABLE 4.1. Typical Room-Temperature Values of Elastic Constants for Isotropic Materials[a]

Material	λ, lb/in.^{2}	μ, lb/in.^{2}	E, lb/in.^{2}	ν
Steels	1.8×10^7	1.2×10^7	3.0×10^7	0.30
Aluminum	7.6×10^6	3.8×10^6	1.0×10^7	0.33
Copper	1.4×10^7	6.8×10^6	1.8×10^7	0.34
Glass	3.6×10^6	3.6×10^6	0.9×10^7	0.25

[a] Note: 1 lb/in.^{2} = 6.90×10^3 N/m^2.

4.9. Governing Equations for Linear Elastic Deformation of an Isotropic Solid

We now collect all of the equations which have thus far been established as applicable to the case of small deformations of an isotropic elastic solid and combine these to form a single set of differential equations governing the displacements.

From Chapter 2, we have first the strain components associated with the small deformation. In terms of the displacement components u_i, these are expressible from equation (2.45) of Chapter 2 as

$$e_{ij} = \frac{1}{2}\left(\frac{\partial u_i}{\partial X_j} + \frac{\partial u_j}{\partial X_i}\right) \tag{4.60}$$

where, it will be remembered, X_i denotes position in the undeformed body.

From Chapter 3, we also have the local form of mass conservation connecting the change in density ϱ' with the initial density ϱ_0 and the displacement gradients $\partial u_i/\partial X_j$ together with the local forms of linear and angular momentum balance connecting the stress components t_{ij} and the body-force components f_i with the motion of the material. These are given from equations (3.56)–(3.58) of Chapter 3 as

$$\varrho = -\varrho_0 \frac{\partial u_i}{\partial X_i} \tag{4.61}$$

$$\varrho_0 \frac{\partial^2 u_i}{\partial t^2} = \frac{\partial t_{ji}}{\partial X_j} + f_i \tag{4.62}$$

$$t_{ij} = t_{ji} \tag{4.63}$$

From the present chapter, we have, finally, the constitutive relations connecting stress and strain components for the case of an isotropic material expressed as

$$t_{ij} = \lambda e_{kk}\,\delta_{ij} + 2\mu e_{ij} \tag{4.64}$$

On combining equations (4.60), (4.62), and (4.64), we thus have the following relation governing the displacement in an isotropic solid:

$$\varrho_0 \frac{\partial^2 u_i}{\partial t^2} = (\lambda + \mu)\frac{\partial^2 u_j}{\partial X_i\,\partial X_j} + \mu \frac{\partial^2 u_i}{\partial X_j\,\partial X_j} + f_i \tag{4.65}$$

The three scalar equations implied by this relation are known as *Navier's equations* and their integration, together with suitable boundary conditions, accordingly determines the displacement components at all points within the solid. If desired, the strain and associated stress can subsequently be established directly from equations (4.60) and (4.64). Similarly, the change in density, if desired, can also be established directly from equation (4.61).

Selected Reading

Malvern, L. E., *Introduction to the Mechanics of a Continuous Medium.* Prentice-Hall, Englewood Cliffs, New Jersey, 1969. Chapter 6 gives a general discussion of constitutive relations. Section 6.7 discusses the principle of material indifference.

Love, A. E. H., *Mathematical Theory of Elasticity.* Cambridge University Press, Cambridge, England, 1927. This is a classical treatise on linear elasticity containing all important developments up to 1927. An excellent historical introduction is included at the beginning.

Frederick, D., and T. S. Chang, *Continuum Mechanics.* Allyn and Bacon, Boston, Massachusetts, 1965. Chapter 5 gives a brief discussion of the basic equations of linear elasticity.

Long, R. R., *Mechanics of Solids and Fluids.* Prentice-Hall, Englewood Cliffs, New Jersey, 1961. The equations of linear elasticity are developed in Chapter 5 and selected problems are considered in Chapter 6.

Leigh, D. C., *Nonlinear Continuum Mechanics.* McGraw-Hill Book Co., New York, 1968. Chapter 8 discusses general principles for constitutive equations, including the principle of material indifference.

Truesdell, C., and W. Noll, "The Non-Linear Field Theories of Mechanics," in *Encyclopedia of Physics*, Vol. III/3. Springer-Verlag, Berlin, 1965. A very advanced treatise on continuum mechanics including detailed development of various constitutive relations. Section 19 provides an interesting discussion of the principle of material indifference.

Exercises

1. For a small static displacement du_i of the surface a of a body of volume v, show in the absence of body forces that the *work* done by the applied surface stress vector t_i is expressible as

$$\oint_a t_i \, du_i \, da = \int_v t_{ij} \, de_{ij} \, dv$$

2. By including the work done by body forces on the left-hand side of the above equation, show that the total work done is still given by the expression on the right-hand side.

3. Show that the condition for the work defined in Exercises 1 and 2 to be independent of the path of straining is that the stress be derivable from a strain-energy function U in accordance with equation (4.35).

4. For a pressure loading such that $t_{11} = t_{22} = t_{33} = -p$, show that the pressure p and the dilatation $\Delta = e_{kk}$ are related for an isotropic solid by

$$p = -K\Delta$$

   where $K = \lambda + \frac{2}{3}\mu$ denotes the *bulk modulus*.

5. Show that the following relations apply:

$$\lambda = \frac{Ev}{(1 + v)(1 - 2v)}$$

$$\mu = \frac{E}{2(1 + v)}$$

$$K = \frac{E}{3(1 - 2v)}$$

   where λ and μ denote the Lamé elastic constants, E and v denote Young's modulus and Poisson's ratio, and K is the bulk modulus as defined in Exercise 4.

6. If an axial stress t_{xx} is applied to a prismatic bar constrained against lateral strain, show that the relation between t_{xx} and e_{xx} is expressible as

$$t_{xx} = \frac{E(1 - v)}{(1 + v)(1 - 2v)} e_{xx}$$

7. Determine the dilatation (change in volume per unit volume) of a cylindrical rod of isotropic elastic material subjected to uniaxial loading.

8. Let u_i denote displacement components satisfying Navier's equation and specified boundary conditions on an elastic body and let u_i' denote different displacement components satisfying Navier's equation and different boundary conditions. Prove that the sum $u_i + u_i'$ is also a solution of Navier's equation satisfying boundary conditions given by the sum of those corresponding to u_i and u_i'. This is the *principle of superposition* applicable to all linear equations.

5

Problems in Elasticity

The previous chapter was concerned with the development of the basic equations governing the small elastic deformation of isotropic solids. Such considerations, though highly restrictive, nevertheless have wide application in engineering analysis. In fact, almost all construction materials exhibit linear elastic behavior for sufficiently small deformations, and a number of these materials can also be regarded as isotropic or nearly so. In the present chapter, we accordingly consider solutions of these basic elasticity equations in connection with certain illustrative problems.

5.1. Longitudinal and Transverse Elastic Waves

We consider the solution of Navier's equations of Chapter 4 for the case where a normal displacement is applied to the face of an otherwise stress-free elastic region. We assume, in particular, the applied boundary displacement to be described, starting from $t = 0$, by the equations

$$u_1 = f(t), \qquad u_2 = u_3 = 0 \tag{5.1}$$

where $f(t)$ denotes some arbitrary function. The problem is illustrated in Figure 5.1.

We may expect in this case that the motion of the solid resulting from the boundary disturbance will be confined to a one-dimensional motion in the direction of the boundary motion. Hence, taking $x = X_1$, and assuming

$$u_1 = u(x, t), \qquad u_2 = u_3 = 0 \tag{5.2}$$

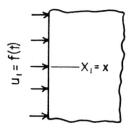

FIGURE 5.1.

we find from Navier's equation,

$$\varrho_0 \frac{\partial^2 u_i}{\partial t^2} = (\lambda + 2\mu) \frac{\partial^2 u_j}{\partial X_i \, \partial X_j} + \mu \frac{\partial^2 u_i}{\partial X_j \, \partial X_i} \tag{5.3}$$

with body forces assumed zero, the following equation:

$$\frac{\partial^2 u}{\partial t^2} = C_1^2 \frac{\partial^2 u}{\partial x^2} \tag{5.4}$$

where C_1 is given by

$$C_1^2 = \frac{\lambda + 2\mu}{\varrho_0} \tag{5.5}$$

The general solution of this equation is expressible in the form

$$u = g(t - x/C_1) + h(t + x/C_1) \tag{5.6}$$

where g and h denote arbitrary functions. The physical interpretation of this solution is straightforward. Consider, for example, the function $g(t - x/C_1)$. At some fixed time t^*, this function will describe a certain displacement distribution along the x direction. Moreover, at the later time $t^* + \Delta t$ and at the forward positions $x + C_1 \Delta t$, direct substitution shows that this same distribution is also described by the function $g(t - x/C_1)$. Thus, this function represents a *wave* of fixed shape traveling in the positive x direction with speed C_1. Similarly, the function $h(t + x/C_1)$ represents a wave of fixed shape traveling in the negative x direction with the same speed C_1.

For the particular problem illustrated in Figure 5.1, the solution of equation (5.4) satisfying the boundary conditions of equation (5.1) and the added initial conditions that $u = \partial u/\partial t = 0$ at time $t = 0$ can be expected to involve waves moving in the positive x direction only. Also, since the waves move at the finite speed C_1, there must necessarily exist at any instant a forward region not yet reached by the wave disturbance. From these considerations, it is thus easily seen that the solution of equation (5.4) satisfying the above boundary and initial conditions is expressible as

$$
\begin{aligned}
u &= f(t - x/C_1), & x \leq C_1 t \\
u &= 0, & x \geq C_1 t
\end{aligned}
\tag{5.7}
$$

Taking as a specific example the case where the boundary condition of equation (5.1) is given by

$$
u_1 = U \sin \sigma t, \qquad u_2 = u_3 = 0
\tag{5.8}
$$

with U and σ denoting constants, we may express the solution to equation (5.4) as

$$
\begin{aligned}
u &= U \sin \sigma(t - x/C_1), & x \leq C_1 t \\
u &= 0, & x \geq C_1 t
\end{aligned}
\tag{5.9}
$$

This solution represents a simple example of a *longitudinal elastic wave* propagating in the x direction with speed C_1, the direction of the resulting particle motion being the same as that of the wave propagation. The displacement distribution corresponding to equation (5.9) is shown in Figure 5.2.

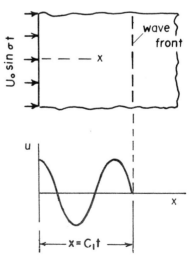

FIGURE 5.2.

The *stress wave* associated with this displacement can easily be determined from the elastic constitutive relations and the strain-displacement relations used in Chapter 4 in deriving Navier's equations. From these relations, we have, in particular, for the longitudinal stress $t_{11} = t_{xx}$ the equation

$$t_{xx} = (\lambda + 2\mu) \frac{\partial u}{\partial x} \tag{5.10}$$

so that the longitudinal stress associated with the above displacement is accordingly expressible as

$$
\begin{aligned}
t_{xx} &= -(\lambda + 2\mu)U(\sigma/C_1) \cos \sigma(t - x/C_1), & x \le C_1 t \\
t_{xx} &= 0, & x > C_1 t
\end{aligned} \tag{5.11}
$$

The maximum stress induced in the material is thus seen to be proportional to the maximum displacement U and to the frequency σ of the disturbance.

It is interesting to note from the above solution that the stress is discontinuous at the wavefront $x = C_1 t$. Its value immediately ahead of this front is, in fact, zero, while its value immediately behind is given by

$$t_{xx} = -(\lambda + 2\mu)U\sigma/C_1$$

Also, if the nonzero boundary condition at $x = 0$ were given by the stress condition

$$t_{xx} = A \cos \sigma t$$

rather than the displacement condition of equation (5.8), then equation (5.11) shows that this same solution would still apply provided U were determined as

$$U = -\frac{AC_1}{(\lambda + 2\mu)\sigma}$$

We note that if a tangential rather than a normal displacement were applied to the boundary of the elastic region considered above, the displacement at interior points would then be expressible as

$$u_2 = v(x, t), \qquad u_1 = u_3 = 0 \tag{5.12}$$

and the solution of Navier's equations analogous to equation (5.7) would then represent a *transverse elastic wave* propagating in the x direction, the direction of the particle motion being in this case perpendicular to the direction of wave propagation and the wave speed C_2 given by

$$C_2^2 = \mu/\varrho_0 \tag{5.13}$$

Application to Bars. The above exact solution may be applied to the study of initial longitudinal wave propagation in bars of uniform rectangular cross section which have suitable lateral stresses t_{yy} and t_{zz} applied in order to prevent lateral motion. The magnitude of these stresses is easily determined from the constitutive relations

$$t_{xx} = (\lambda + 2\mu)e_{xx}, \qquad t_{yy} = t_{zz} = \lambda e_{xx} \tag{5.14}$$

which follow from the general elastic relations when the lateral strains are set equal to zero. Solving these equations for the lateral stresses in terms of the longitudinal stress t_{xx}, we find

$$t_{yy} = t_{zz} = \frac{\lambda}{\lambda + 2\mu} t_{xx} \tag{5.15}$$

In the case of bars having stress-free lateral surfaces, an exact elasticity solution for wave disturbances is far more difficult than that described above because the accompanying lateral motion gives rise to a nonuniform stress distribution over the cross section of the bar. However, since this nonuniformity must vanish with the vanishing of the lateral dimensions of the bar, an approximate treatment can be devised which ignores the nonuniformity altogether yet applies for those practical cases where the lateral dimensions of the bar may be considered small in comparison with the characteristic length of the wave disturbance.

According to this theory, which applies to bars of uniform but otherwise arbitrary cross sections, the only nonzero stress in the bar is the longitudinal stress t_{xx}, so that the momentum equation in the longitudinal x direction is expressible simply as

$$\varrho_0 \frac{\partial^2 u}{\partial t^2} = \frac{\partial t_{xx}}{\partial x} \tag{5.16}$$

The elastic constitutive relation for this case reduces to

$$t_{xx} = Ee_{xx} = E\frac{\partial u}{\partial x} \tag{5.17}$$

where E denotes Young's modulus. On combining these two relations, we can express Navier's equation for this case as

$$\frac{\partial^2 u}{\partial t^2} = C_0^2 \frac{\partial^2 u}{\partial x^2} \tag{5.18}$$

where C_0 is given by

$$C_0^2 = E/\varrho_0 \tag{5.19}$$

This equation is identical to that of equation (5.4) except that the *bar wave speed* C_0 replaces the longitudinal wave speed C_1. The corresponding solution to the problem of a long bar having, for example, the longitudinal displacement

$$u = U \sin \sigma t \tag{5.20}$$

applied to one end is therefore determined immediately from equation (5.9) as

$$
\begin{aligned}
u &= U \sin \sigma(t - x/C_0), & x &\le C_0 t \\
u &= 0, & x &\ge C_0 t
\end{aligned}
\tag{5.21}
$$

This solution is sketched in Figure 5.3.

The stress in the bar may be determined from equation (5.17) as

$$
\begin{aligned}
t_{xx} &= -EU(\sigma/C_0) \cos \sigma(t - x/C_0), & x &\le C_0 t \\
t_{xx} &= 0, & x &> C_0 t
\end{aligned}
\tag{5.22}
$$

Wave Reflections and Transmissions.   In connection with the general problem of waves in solids, it is perhaps instructive to consider briefly the situation for two bars of different material welded together as in Figure 5.4. When an incident wave in bar I encounters the interface between the two bars, a part of this wave must be transmitted and a part reflected in order to satisfy the displacement and stress boundary conditions at the interface.

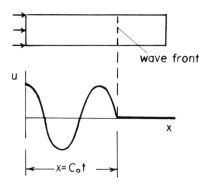

FIGURE 5.3.

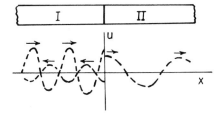

FIGURE 5.4.

As an example, consider the following incident (0), reflected (R), and transmitted (T) waves:

$$u_0 = U_0 \sin \sigma_0(t - x/C_I)$$
$$u_R = U_R \sin \sigma_R(t + x/C_I) \tag{5.23}$$
$$u_T = U_T \sin \sigma_T(t - x/C_{II})$$

where x is measured from the interface and C_I and C_{II} denote values of the bar wave speed in the two materials.

The boundary conditions at the interface $x = 0$ require that the displacement and stress be continuous there. From the above waveforms, we thus require that

$$u_0 = u_R + u_T \tag{5.24}$$

and

$$E_I\left(\frac{\partial u_0}{\partial x} + \frac{\partial u_R}{\partial x}\right) = E_{II}\frac{\partial u_T}{\partial x} \tag{5.25}$$

where E_I and E_{II} denote Young's modulus in the two materials. Substituting the above wave solutions into these relations and simplifying, we find

$$\sigma_0 = \sigma_R = \sigma_T \tag{5.26}$$

$$U_R = \frac{\varrho_I C_I - \varrho_{II} C_{II}}{\varrho_I C_I + \varrho_{II} C_{II}} U_0 \tag{5.27}$$

$$U_T = \frac{2\varrho_I C_I}{\varrho_I C_I + \varrho_{II} C_{II}} U_0 \tag{5.28}$$

where ϱ_I and ϱ_{II} denote the densities of the two materials. The total displacement and stress at the interface is thus expressible in terms of the incident displacement u_0 and the incident stress t_{xx}^0 as

$$u = \frac{2\varrho_I C_I}{\varrho_I C_I + \varrho_{II} C_{II}} u_0 \tag{5.29}$$

$$t_{xx} = \frac{2\varrho_{II} C_{II}}{\varrho_I C_I + \varrho_{II} C_{II}} t_{xx}^0 \tag{5.30}$$

In the special idealized case where the material II is rigid, the wave speed C_{II} is unbounded, and these equations then show that the displacement is zero at the interface and that the maximum stress is twice that of the incident wave. Also, in the case where the interface is a free surface, $\varrho_{II} C_{II} = 0$ and the equations then show that the stress is zero at the free surface and that the displacement is twice that of the incident wave.

5.2. Static Twisting of Rods and Bars

We next consider the static twisting of a uniform rod or bar of isotropic material by torques applied at its ends. We choose one end to be fixed against rotation and take coordinates $x = X_1$, $y = X_2$, $z = X_3$ as shown in Figure 5.5. If we assume tentatively that cross sections of the bar simply rotate and simultaneously warp during the twisting, we may take the displacements $u_1 = u$, $u_2 = v$, $u_3 = w$ to be expressible as

$$u = -\alpha z y, \qquad v = \alpha z x, \qquad w = \alpha \psi(x, y) \tag{5.31}$$

where $\psi(x, y)$ denotes Saint-Venant's *warping function* and αz denotes the angle of rotation of the cross section a distance z from the fixed end.

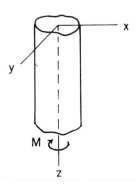

FIGURE 5.5.

On substituting equation (5.31) into Navier's equation, as given by equation (4.65) of Chapter 4, we find that this equation is satisfied for zero body forces provided the warping function satisfies

$$\frac{\partial^2 \psi}{\partial x^2} + \frac{\partial^2 \psi}{\partial y^2} = 0 \tag{5.32}$$

From the constitutive relations and strain-displacement relations given by equations (4.60) and (4.64) of Chapter 4, we find further that the stresses associated with the above displacements are expressible as

$$t_{xx} = t_{yy} = t_{zz} = t_{xy} = 0$$

$$t_{xz} = \mu \alpha \left(\frac{\partial \psi}{\partial x} - y \right) \tag{5.33}$$

$$t_{yz} = \mu \alpha \left(\frac{\partial \psi}{\partial y} + x \right)$$

On the force-free lateral surface of the bar, the outward normal has components $(n_x, n_y, 0)$ and the boundary conditions given by equation (3.47) of Chapter 3 thus require that

$$\hat{t}_x = t_{xx} n_x + t_{xy} n_y = 0$$
$$\hat{t}_y = t_{xy} n_x + t_{yy} n_y = 0 \tag{5.34}$$
$$\hat{t}_z = t_{xz} n_x + t_{yz} n_y = 0$$

In view of the above stress components, the first two of these equations are seen to be satisfied identically and the third reduces to

$$\left(\frac{\partial \psi}{\partial x} - y \right) n_x + \left(\frac{\partial \psi}{\partial y} + x \right) n_y = 0 \tag{5.35}$$

Since the nonzero stress components t_{xz} and t_{yz} of equation (5.33) produce x and y components of force acting on an element of area dA of the cross section, we must finally require that these stresses produce the applied end torque M with no resultant forces in the x and y directions. These conditions can be expressed as

$$M = \int_A (x t_{yz} - y t_{xz}) \, dA$$

$$F_x = \int_A t_{xz} \, dA = 0 \tag{5.36}$$

$$F_y = \int_A t_{yz} \, dA = 0$$

Circular Rod. In the case of a rod of circular cross section, the above conditions can all be seen to be satisfied by requiring that the warping function ψ vanish everywhere. The assumed displacements of equation (5.31) are thus those of the correct elasticity solution for this case. Cross sections of the rod are accordingly seen to remain plane and simply rotate during the twisting.

From the first of equations (5.36) we find, in particular, that

$$M = J\mu\alpha \tag{5.37}$$

where $J = \pi R^4/2$, R denoting the radius of the rod. When the twisting moment is specified, equation (5.37) may therefore be used to determine the angle of rotation αz at any section of the rod. Combining this result with the stress relations of equation (5.33), we may also express the stresses in terms of the twisting moment as

$$t_{xz} = -\frac{M}{J}\,y, \qquad t_{yz} = \frac{M}{J}\,x \tag{5.38}$$

The maximum stress is thus seen to occur at $x = 0$, $y = R$ or $x = R$, $y = 0$ and is given by

$$t_{\max} = MR/J \tag{5.39}$$

It is worth noting that the above solution can also be applied to the twisting of a hollow circular rod having inside and outside radii R_i and R_0, respectively, provided, from equation (5.36), that $J = \pi(R_0^4 - R_i^4)/2$.

Rectangular Bar. The solution for the twisting of a bar of rectangular section is not so simple as that for a circular rod. Taking the cross-section dimensions to be as shown in Figure 5.6, and introducing the variable ϕ such that

$$\psi = xy + \phi \tag{5.40}$$

we find that equation (5.32) becomes

$$\frac{\partial^2\phi}{\partial x^2} + \frac{\partial^2\phi}{\partial y^2} = 0 \tag{5.41}$$

and the boundary condition of equation (5.35) becomes

$$\frac{\partial\phi}{\partial x} = 0 \qquad \text{on} \quad x = \pm a$$

$$\frac{\partial\phi}{\partial y} = -2x \qquad \text{on} \quad y = \pm b \tag{5.42}$$

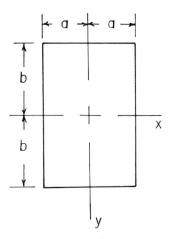

FIGURE 5.6.

With these conditions we may choose a solution of equation (5.41) in the form

$$\phi = A \sin kx \sinh ky \tag{5.43}$$

where the above boundary condition on $x = \pm a$ requires

$$A \cos ka = 0$$

or

$$k = \frac{n\pi}{2a}, \qquad n = 1, 3, 5, \ldots \tag{5.44}$$

For each of these values of n, we have a corresponding solution for ϕ, so that, on taking all values of n, we thus have

$$\phi = \sum_{n=1,3,5}^{\infty} A_n \sin\left(\frac{n\pi}{2a} x\right) \sinh\left(\frac{n\pi}{2a} y\right) \tag{5.45}$$

The boundary condition on $y = \pm b$ requires further that

$$-2x = \sum_{n=1,3,5}^{\infty} A_n\left(\frac{n\pi}{2a}\right) \cosh\left(\frac{n\pi}{2a} b\right) \sin\left(\frac{n\pi}{2a} x\right)$$

From the theory of Fourier series, this condition can be satisfied by taking the coefficients A_n in the form

$$A_n = (-1)^{(n+1)/2} \frac{32a^2}{n^3\pi^3} \frac{1}{\cosh(n\pi b/2a)}$$

Hence, the solution for ϕ is thus expressible in final form as

$$\phi = \frac{32a^2}{\pi^3} \sum_{n=1,3,5}^{\infty} (-1)^{(n+1)/2} \frac{1}{n^3} \frac{\sinh(n\pi y/2a)}{\cosh(n\pi b/2a)} \sin\frac{n\pi x}{2a} \tag{5.46}$$

It remains to examine the resultant force and moment boundary conditions as prescribed by equations (5.36). Substituting the stress relation of equation (5.33) into the second of these, we find, in particular, that

$$F_x = \mu\alpha \int_{-b}^{b} \int_{-a}^{a} \left(\frac{\partial\psi}{\partial x} - y\right) dx\, dy \tag{5.47}$$

which integrates directly to

$$F_x = \mu\alpha \int_{-b}^{b} [\psi(a, y) - \psi(-a, y)]\, dy \tag{5.48}$$

By examining equations (5.40) and (5.46), it can easily be seen that the integral of ψ with respect to y will be an even function of y, so that, on performing the integration of equation (5.48) over the indicated limits, the resultant force will thus vanish as required.

A similar argument likewise shows that $F_y = 0$, in agreement with the requirements of the third of equations (5.36).

The resultant moment condition at any section, as given by the first of equations (5.36) may finally be expressed with the help of equation (5.33) as

$$M = \mu\alpha \int_{-b}^{b} \int_{-a}^{a} \left(x^2 + y^2 + x\frac{\partial\psi}{\partial y} - y\frac{\partial\psi}{\partial x}\right) dx\, dy \tag{5.49}$$

All of the above results, taken together, thus show that the assumed displacement components of equation (5.31) are, in fact, consistent with the correct elasticity solution for the twisting of a bar of rectangular section. In particular, with the warping function given by equations (5.40) and (5.46), we may determine the nonzero stress components and the moment-twist relation as

$$t_{xz} = \frac{16\mu\alpha a}{\pi^2} \sum_{n=1,3,5}^{\infty} (-1)^{(n+1)/2} \frac{1}{n^2} \frac{\sinh(n\pi y/2a)}{\cosh(n\pi b/2a)} \cos\frac{n\pi x}{2a} \tag{5.50}$$

$$t_{yz} = 2\mu\alpha x + \frac{16\mu\alpha a}{\pi^2} \sum_{n=1,3,5}^{\infty} (-1)^{(n+1)/2} \frac{1}{n^2} \frac{\cosh(n\pi y/2a)}{\cosh(n\pi b/2a)} \sin\frac{n\pi x}{2a} \tag{5.51}$$

$$M = \frac{16}{3}\mu\alpha a^3 b - \left(\frac{4}{\pi}\right)^5 \mu\alpha a^4 \sum_{n=1,3,5}^{\infty} \frac{1}{n^5} \tanh\frac{n\pi b}{2a} \tag{5.52}$$

TABLE 5.1. Constants for Torsion of Rectangular Bars

b/a	K_1	K_2	b/a	K_1	K_2
1.0	0.142	0.675	5.0	0.292	0.999
1.5	0.196	0.848	6.0	0.299	1.000
2.0	0.229	0.930	7.0	0.304	1.000
2.5	0.250	0.968	8.0	0.307	1.000
3.0	0.264	0.985	9.0	0.310	1.000
3.5	0.274	0.993	10.0	0.312	1.000
4.0	0.281	0.997	∞	0.333	1.000

The above series solutions converge rapidly and numerical results can be worked out without difficulty. By direct calculation, it can be found that the maximum stress occurs at the midpoints of the longer sides of the cross section; that is, at $x = \pm a$, $y = 0$. If we express the moment-twist and maximum stress-twist relations as

$$M = K_1(2a)^3(2b)\mu\alpha \tag{5.53}$$

and

$$t_{\max} = K_2(2a)\mu\alpha \tag{5.54}$$

we can thus use the above solutions to calculate the coefficients K_1 and K_2 for various ratios b/a. Some typical results are shown in Table 5.1.

5.3. Saint-Venant's Principle

The solutions just obtained for the twisting of rods of circular and rectangular section provide a convenient means for introducing *Saint-Venant's principle* regarding the application of static elasticity solutions to technically important problems. If, for example, we examine the stress components acting at the end of the circular rod where an external torque is applied, we find that they must be distributed according to the results of equation (5.38). The actual application of such a precise stress distribution to obtain the resultant twisting moment would, however, clearly be difficult to realize in practice. If the precise satisfaction of this end-stress distribution were necessary in order to make the details of the above elasticity solution valid all along the rod, its application to technically important problems involving the twisting of rods would therefore be extremely limited. Fortunately, however, intuition suggests and experiment confirms that the elastic behavior of the rod at some distance, say a diameter or two, from its ends is insensitive to the manner in which the surface stress distributions are actually applied, provided they simply give rise to a resultant moment with no resultant forces in accordance with equations (5.36).

The above observation, when generalized to other static elasticity solutions, accordingly yields Saint-Venant's principle in the following form: If a system of forces acting on a part of the surface of an elastic solid is replaced by a different system of forces which retain the same resultant forces and moments, the elasticity solutions of the two problems will not differ sensibily from one another at distances which are large in comparison with the linear dimensions of the part of the surface in question.

5.4. Compatibility Equations

In obtaining solutions to the problems of the previous sections, the strategy followed involved the assumption of tentative displacement components and a subsequent examination of their consistency with Navier's equations and the boundary and initial conditions. In certain equilibrium problems, it is, however, often easier to make tentative assumptions about, or otherwise deal with, the stress or strain components rather than the displacement components and subsequently examine their consistency with imposed boundary conditions and the governing elasticity equations. Since assumptions of stress components necessarily imply through the constitutive relations corresponding assumptions about the strain components, it is clear that, in following this latter method, great care must be taken in order to ensure that the six assumed or implied strain components are, in fact, derivable from only three displacement components in accordance with the strain-displacement relations. The mathematical restrictions on the strain components which ensure that this is the case are known as the *compatibility equations*.

To establish these equations, let us first suppose that displacement components do indeed exist. In this case, we have from Chapter 2 that the strain components e_{ij} and the rotation components w_{ij} are given by

$$e_{ij} = \frac{1}{2} \left(\frac{\partial u_i}{\partial X_j} + \frac{\partial u_j}{\partial X_i} \right) \tag{5.55}$$

$$w_{ij} = \frac{1}{2} \left(\frac{\partial u_i}{\partial X_j} - \frac{\partial u_j}{\partial X_i} \right) \tag{5.56}$$

On adding these equations, we have

$$\frac{\partial u_i}{\partial X_j} = e_{ij} + w_{ij} \qquad (5.57)$$

From this equation, we have by differentiation

$$\frac{\partial^2 u_i}{\partial X_k \, \partial X_j} = \frac{\partial e_{ij}}{\partial X_k} + \frac{\partial w_{ij}}{\partial X_k} \qquad (5.58)$$

Hence, with

$$\frac{\partial^2 u_i}{\partial X_k \, \partial X_j} = \frac{\partial^2 u_i}{\partial X_j \, \partial X_k} \qquad (5.59)$$

we have from equation (5.58) that

$$\frac{\partial e_{ij}}{\partial X_k} - \frac{\partial e_{ik}}{\partial X_j} = \frac{\partial w_{ik}}{\partial X_j} - \frac{\partial w_{ij}}{\partial X_k} \qquad (5.60)$$

But from equation (5.56) we also have

$$\frac{\partial w_{ik}}{\partial X_j} - \frac{\partial w_{ij}}{\partial X_k} = \frac{\partial w_{jk}}{\partial X_i} \qquad (5.61)$$

so that equation (5.60) becomes expressible as

$$\frac{\partial e_{ij}}{\partial X_k} - \frac{\partial e_{ik}}{\partial X_j} = \frac{\partial w_{jk}}{\partial X_i} \qquad (5.62)$$

On differentiating this last equation with respect to X_l and using

$$\frac{\partial^2 w_{jk}}{\partial X_l \, \partial X_i} = \frac{\partial^2 w_{jk}}{\partial X_i \, \partial X_l} \qquad (5.63)$$

we have, finally, the following compatibility equations:

$$\frac{\partial^2 e_{ij}}{\partial X_k \, \partial X_l} \frac{\partial^2 e_{lk}}{\partial X_j \, \partial X_i} - \frac{\partial^2 e_{ik}}{\partial X_j \, \partial X_l} - \frac{\partial^2 e_{lj}}{\partial X_k \, \partial X_i} = 0 \qquad (5.64)$$

The above argument shows that equation (5.64) is *necessary* if displacement components exist that give rise to strain components in accordance with equation (5.55). To show that it is also *sufficient* to ensure that they exist, we must take this equation as our starting point and reverse the argument. Thus, we rewrite equation (5.64) in the following form:

$$\frac{\partial}{\partial X_l}\left(\frac{\partial e_{ij}}{\partial X_k} - \frac{\partial e_{ik}}{\partial X_j}\right) + \frac{\partial}{\partial X_i}\left(\frac{\partial e_{lk}}{\partial X_j} - \frac{\partial e_{lj}}{\partial X_k}\right) = 0 \qquad (5.65)$$

By inspection, this equation is seen to imply the existence of antisymmetric tensor components r_{ij} such that

$$\frac{\partial r_{jk}}{\partial X_i} = \frac{\partial e_{ij}}{\partial X_k} - \frac{\partial e_{ik}}{\partial X_j} \qquad (5.66)$$

Taking r_{ij} in the form

$$r_{ij} = \frac{1}{2}\left(\frac{\partial f_i}{\partial X_j} - \frac{\partial f_j}{\partial X_i}\right) \qquad (5.67)$$

where f_i denotes components of a vector, equation (5.66) becomes expressible as

$$\frac{\partial}{\partial X_k}\left(2e_{ij} - \frac{\partial f_j}{\partial X_i}\right) = \frac{\partial}{\partial X_j}\left(2e_{ik} - \frac{\partial f_k}{\partial X_i}\right) \qquad (5.68)$$

Again, by inspection, this equation is seen to imply the existence of vector components g_i such that

$$2e_{ij} - \frac{\partial f_j}{\partial X_i} = \frac{\partial g_i}{\partial X_j} \qquad (5.69)$$

Hence, choosing $g_i = f_i = u_i$, we have finally that e_{ij} may be expressed as

$$e_{ij} = \frac{1}{2}\left(\frac{\partial u_i}{\partial X_j} + \frac{\partial u_j}{\partial X_i}\right) \qquad (5.70)$$

so that the compatibility equations themselves are sufficient to ensure the existence of displacement components connected with the strain components by equation (5.55).

We note that, of the 81 scalar equations implied by the compatibility relation of equation (5.64), only six such equations are actually distinct. These may be written as

$$\frac{\partial^2 e_{11}}{\partial X_2\,\partial X_3} = \frac{\partial^2 e_{12}}{\partial X_1\,\partial X_3} - \frac{\partial^2 e_{23}}{\partial X_1^2} + \frac{\partial^2 e_{31}}{\partial X_1\,\partial X_2} \qquad (5.71)$$

$$\frac{\partial^2 e_{22}}{\partial X_1\,\partial X_3} = \frac{\partial^2 e_{23}}{\partial X_2\,\partial X_1} - \frac{\partial^2 e_{31}}{\partial X_2^2} + \frac{\partial^2 e_{12}}{\partial X_2\,\partial X_3} \qquad (5.72)$$

$$\frac{\partial^2 e_{33}}{\partial X_1\, \partial X_2} = \frac{\partial^2 e_{31}}{\partial X_3\, \partial X_2} - \frac{\partial^2 e_{12}}{\partial X_3^2} + \frac{\partial^2 e_{23}}{\partial X_3\, \partial X_1} \tag{5.73}$$

$$2\frac{\partial^2 e_{12}}{\partial X_1\, \partial X_2} = \frac{\partial^2 e_{11}}{\partial X_2^2} + \frac{\partial^2 e_{22}}{\partial X_1^2} \tag{5.74}$$

$$2\frac{\partial^2 e_{23}}{\partial X_2\, \partial X_3} = \frac{\partial^2 e_{22}}{\partial X_3^2} + \frac{\partial^2 e_{33}}{\partial X_2^2} \tag{5.75}$$

$$2\frac{\partial^2 e_{31}}{\partial X_3\, \partial X_1} = \frac{\partial^2 e_{33}}{\partial X_1^2} + \frac{\partial^2 e_{11}}{\partial X_3^2} \tag{5.76}$$

It is worth noting also in connection with these six compatibility equations that, since they were derived from only three independent displacement components, they represent, in fact, only three mathematically independent equations.

5.5. Plane Strain and Plane Stress

When an elastic body is subjected to forces in such a way that the strain components, say e_{xz}, e_{yz}, e_{zz}, referred to axes $X_1 = x$, $X_2 = y$, $X_3 = z$ satisfy

$$e_{xz} = e_{yz} = e_{zz} = 0 \tag{5.77}$$

and the remaining strain components are functions of x and y only, we have a state of *plane strain* existing in the solid. Such a strained state can be realized, for example, in a long cylindrical body having applied loads which act normal to the axis of the cylinder and which are uniform along its length as indicated in Figure 5.7.

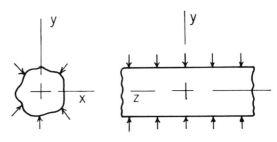

FIGURE 5.7.

For the case of plane strain, we have from the isotropic constitutive relations of equation (4.59) of Chapter 4 that

$$e_{xz} = \frac{1+\nu}{E} t_{xz} = 0 \tag{5.78}$$

$$e_{yz} = \frac{1+\nu}{E} t_{yz} = 0 \tag{5.79}$$

$$e_{zz} = \frac{1}{E} t_{zz} - \frac{\nu}{E} (t_{xx} + t_{yy}) = 0 \tag{5.80}$$

so that

$$t_{yz} = t_{xz} = 0 \tag{5.81}$$

and

$$t_{zz} = \nu(t_{xx} + t_{yy}) \tag{5.82}$$

Using this last equation, we can express the remaining constitutive relations as

$$e_{xx} = \frac{1+\nu}{E} [(1-\nu)t_{xx} - \nu t_{yy}] \tag{5.83}$$

$$e_{yy} = \frac{1+\nu}{E} [(1-\nu)t_{yy} - \nu t_{xx}] \tag{5.84}$$

$$e_{xy} = \frac{1+\nu}{E} t_{xy} \tag{5.85}$$

Remembering that the strain components e_{xx}, e_{yy}, and e_{xy} are functions of x and y only, it can be seen from the first two of these equations that the stress components t_{xx} and t_{yy} must likewise be functions of x and y only. In view of equations (5.82) and (5.85), the same is also seen to be true for t_{zz} and t_{xy}.

In contrast with plane-strain loadings, we may consider, alternatively, forces applied to an elastic body in such a way that the stress components, say t_{xz}, t_{yz}, t_{zz}, referred to axes $X_1 = x$, $X_2 = y$, $X_3 = z$ satisfy

$$t_{xz} = t_{yz} = t_{zz} = 0 \tag{5.86}$$

In this case, we then have a state of *plane stress* existing in the solid. Such a stress state can, for example, be obtained in a thin plate or beam having loads applied only along its edges as shown in Figure 5.8. Under these

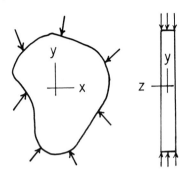

FIGURE 5.8.

circumstances, the isotropic constitutive relations of Chapter 4 can be written simply as

$$e_{xx} = \frac{1}{E} t_{xx} - \frac{v}{E} t_{yy} \qquad (5.87)$$

$$e_{yy} = \frac{1}{E} t_{yy} - \frac{v}{E} t_{xx} \qquad (5.88)$$

$$e_{zz} = -\frac{v}{E} (t_{xx} + t_{yy}) \qquad (5.89)$$

$$e_{xy} = \frac{1+v}{E} t_{xy} \qquad (5.90)$$

$$e_{xz} = e_{yz} = 0 \qquad (5.91)$$

In the absence of body forces, the equilibrium (momentum) equations for either plane strain or plane stress are expressible from equation (3.34) of Chapter 3 as

$$\frac{\partial t_{xx}}{\partial x} + \frac{\partial t_{xy}}{\partial y} = 0 \qquad (5.92)$$

$$\frac{\partial t_{xy}}{\partial x} + \frac{\partial t_{yy}}{\partial y} = 0 \qquad (5.93)$$

These equations can be satisfied automatically by the introduction of the so-called *Airy stress function* such that

$$t_{xx} = \frac{\partial^2 \phi}{\partial y^2}, \qquad t_{yy} = \frac{\partial^2 \phi}{\partial x^2}, \qquad t_{xy} = -\frac{\partial^2 \phi}{\partial x \, \partial y} \qquad (5.94)$$

It remains to consider the compatibility relations. From equation (5.74) we have in particular

$$\frac{\partial^2 e_{xx}}{\partial y^2} + \frac{\partial^2 e_{yy}}{\partial x^2} = 2\frac{\partial^2 e_{xy}}{\partial x\,\partial y} \qquad (5.95)$$

On combining the above plane-strain or plane-stress constitutive relations with the stress-function definitions and this compatibility equation, we find for either case that ϕ must satisfy

$$\frac{\partial^4 \phi}{\partial x^4} + 2\frac{\partial^4 \phi}{\partial x^2\,\partial y^2} + \frac{\partial^4 \phi}{\partial y^4} = 0 \qquad (5.96)$$

We need finally to consider the remaining compatibility equations. In the case of plane strain, these may easily be seen to be identically satisfied by the vanishing of the strain components e_{xz}, e_{yz}, and e_{zz} and by the dependence of the remaining strain components on the coordinates x and y only. However, the same is no longer true for the case of plane stress where e_{zz} does not vanish and where the remaining nonzero strain components are not necessarily required to depend only on the coordinates x and y.

What the above means, of course, is that the strains and stresses in plane stress must, in general, depend not only on the in-plane coordinates x and y but also on the normal coordinate z. In this case, the stress function ϕ must itself vary with both the in-plane coordinates x and y, as required by equation (5.96), and also with the normal coordinate z as required by the compatibility equations. For thin plates and beams, this latter variation will, however, necessarily be small and may be neglected without committing serious error. Under this restriction, solutions to plane-stress problems thus reduce simply to considerations of equation (5.96) and, accordingly, parallel in mathematical structure the case of plane strain.

5.6. Bending of a Thin Beam by Uniform Loading

As an example of the above considerations, we consider the case of a simply supported thin beam of rectangular section subjected to uniform loading p per unit length as shown in Figure 5.9. Such a problem can clearly be regarded as a plane-stress problem with in-plane coordinates x and y as indicated in the figure. The governing plane stress equation for the stress function is given from equation (5.96) as

$$\frac{\partial^4 \phi}{\partial x^4} + 2\frac{\partial^4 \phi}{\partial x^2\,\partial y^2} + \frac{\partial^4 \phi}{\partial y^4} = 0 \qquad (5.97)$$

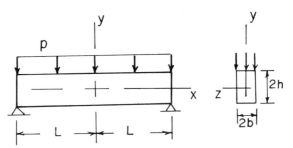

FIGURE 5.9.

and the boundary conditions on the problem require that at $y = +h$

$$\hat{t}_x = t_{xy} = 0$$
$$\hat{t}_y = t_{yy} = -p/2b$$

(5.98)

and that at $y = -h$

$$\hat{t}_x = t_{xy} = 0$$
$$\hat{t}_y = t_{yy} = 0$$

(5.99)

Also, at the ends $x = \pm L$, we require with the help of Saint-Venant's principle that

$$2b \int_{-c}^{+c} t_{xy}\, dy = \pm pL$$

$$2b \int_{-c}^{+c} t_{xx}\, dy = 0$$

(5.100)

$$2b \int_{-c}^{+c} yt_{xx}\, dy = 0$$

In view of the boundary conditions given by equations (5.98) and (5.99), we may attempt a stress-function solution of equation (5.97) corresponding to the case where the stress component t_{yy} is independent of x and where the stress function ϕ is symmetric about $x = 0$. From the stress-function definitions of equation (5.94), this accordingly suggests an expression for ϕ of the form

$$\phi = x^2\phi_1(y) + \phi_2(y)$$

(5.101)

where ϕ_1 and ϕ_2 denote functions of y only.

Substituting this expression into equation (5.97), we find

$$4 \frac{d^2\phi_1}{dy^2} + x^2 \frac{d^4\phi_1}{dy^4} + \frac{d^4\phi_2}{dy^4} = 0 \tag{5.102}$$

Since ϕ_1 and ϕ_2 are functions of y only, we must accordingly choose ϕ_1 such that

$$\frac{d^4\phi_1}{dy^4} = 0 \tag{5.103}$$

Noticing further from the first of the boundary conditions given by equations (5.98) and (5.99) that ϕ_1 must be an odd function of y, we therefore have the solution for ϕ_1 in the form

$$\phi_1 = a + cy + ey^3 \tag{5.104}$$

where a, c, and e denote constants.

From equation (5.102), we then have also that

$$\frac{d^4\phi_2}{dy^4} = -24ey \tag{5.105}$$

so that ϕ_2 is given as

$$\phi_2 = -\frac{e}{5} y^5 + \frac{f}{6} y^3 + \frac{g}{2} y^2 \tag{5.106}$$

where f and g denote integration constants and where linear and constant terms have been ignored since they do not affect the stress distribution.

With the above functions, we have the stress components given from equations (5.94) and (5.101) as

$$t_{xx} = 2ey(3x^2 - y^2) + fy + g$$
$$t_{yy} = 2(a + cy + ey^3) \tag{5.107}$$
$$t_{xy} = -2x(c + 3ey^2)$$

On substituting these expressions into the boundary conditions given by equations (5.98) and (5.99), it is easily established that all are satisfied provided the constants are taken as

$$a = -\frac{p}{8b}, \qquad c = -\frac{3p}{16bh}, \qquad e = \frac{p}{16bh^3}$$

In addition to these values, the boundary conditions of equations (5.100) are also easily seen to be satisfied provided the remaining two constants take the values

$$f = \frac{3}{8} \frac{p}{bh} \left(\frac{2}{5} - \frac{L^2}{h^2} \right), \qquad g = 0$$

Combining the above expressions, we thus find that the stress components are given as

$$t_{xx} = \frac{p}{2I} (x^2 - L^2)y - \frac{p}{2I} \left(\frac{2}{3} y^3 - \frac{2}{5} h^2 y \right)$$

$$t_{yy} = \frac{p}{2I} \left(\frac{1}{3} y^3 - h^2 y - \frac{2}{3} h^3 \right) \qquad\qquad (5.108)$$

$$t_{xy} = \frac{p}{2I} (h^2 - y^2)x$$

where I has been written in place of $4bh^3/3$.

If desired, the displacements associated with this stress distribution can be calculated by integrating the corresponding strain components.

5.7. Equations for Plane Strain and Plane Stress in Polar Coordinates

In solving plane strain and plane stress problems involving circular boundaries, it is generally more convenient to use cylindrical polar coordinates (r, θ, z) rather than rectangular coordinates (x, y, z) as the independent position variables. These coordinates have been discussed previously in Chapters 2 and 3 and are illustrated in Figure 5.10 in the x–y plane.

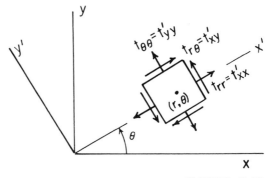

FIGURE 5.10.

From Chapters 2 and 3, we recall that the stress and strain components t_{rr}, $t_{r\theta}$, e_{rr}, etc., in polar coordinates are defined simply as the rectangular components t'_{xx}, t'_{xy}, e'_{xx}, etc. referred to axes x', y', z' with x' and y' in the directions of increasing r and θ and z' in the direction of z, as indicated in Figure 5.10. For the case of plane strain in polar coordinates, we may, accordingly, assume that the strain components e_{rz}, $e_{\theta z}$, and e_{zz} satisfy

$$e_{rz} = e_{\theta z} = e_{zz} = 0 \tag{5.109}$$

and that the remaining components are functions of r and θ only. Similarly, for the case of plane stress we may assume that the stress components t_{rz}, $t_{\theta z}$, and t_{zz} satisfy

$$t_{rz} = t_{\theta z} = t_{zz} = 0 \tag{5.110}$$

For both of these cases, the governing equation for the stress function ϕ, as given by equation (5.96), may be written in rectangular coordinates as

$$\left(\frac{\partial^2}{\partial x^2} + \frac{\partial^2}{\partial y^2} \right)\left(\frac{\partial^2 \phi}{\partial x^2} + \frac{\partial^2 \phi}{\partial y^2} \right) = 0 \tag{5.111}$$

Using the transformation procedures of Chapters 2 and 3, we find this equation expressible in polar coordinates as

$$\left(\frac{\partial^2}{\partial r^2} + \frac{1}{r} \frac{\partial}{\partial r} + \frac{1}{r^2} \frac{\partial^2}{\partial \theta^2} \right)\left(\frac{\partial^2 \phi}{\partial r^2} + \frac{1}{r} \frac{\partial \phi}{\partial r} + \frac{1}{r^2} \frac{\partial^2 \phi}{\partial \theta^2} \right) = 0 \tag{5.112}$$

The stress function definitions of equation (5.94) are likewise expressible in polar coordinates as

$$t_{rr} = \frac{1}{r} \frac{\partial \phi}{\partial r} + \frac{1}{r^2} \frac{\partial^2 \phi}{\partial \theta^2}$$

$$t_{\theta\theta} = \frac{\partial^2 \phi}{\partial r^2} \tag{5.113}$$

$$t_{r\theta} = \frac{1}{r^2} \frac{\partial \phi}{\partial \theta} - \frac{1}{r} \frac{\partial^2 \phi}{\partial r \, \partial \theta}$$

The constitutive relations relating the polar strain components to the above stress components for the case of plane strain are determined from equations (5.83)–(5.85) as

$$e_{rr} = \frac{1+\nu}{E} [(1-\nu)t_{rr} - \nu t_{\theta\theta}]$$

$$e_{\theta\theta} = \frac{1+\nu}{E} [(1-\nu)t_{\theta\theta} - \nu t_{rr}] \qquad (5.114)$$

$$e_{r\theta} = \frac{1+\nu}{E} t_{r\theta}$$

For the case of plane stress, we have, similarly, from equations (5.87)–(5.90) that

$$e_{rr} = \frac{1}{E} (t_{rr} - \nu t_{\theta\theta})$$

$$e_{\theta\theta} = \frac{1}{E} (t_{\theta\theta} - \nu t_{rr}) \qquad (5.115)$$

$$e_{r\theta} = \frac{1+\nu}{E} t_{r\theta}$$

5.8. Thick-Walled Cylinder under Internal Pressure

We illustrate the above equations in polar coordinates by considering the problem of a long, hollow circular cylinder of isotropic elastic material subjected to an internal pressure loading p as shown in Figure 5.11.

The boundary conditions on the problem are expressible at the inside radius $r = a$ as

$$\hat{t}_r = t_{rr}n_r + t_{r\theta}n_\theta = p$$
$$\hat{t}_\theta = t_{r\theta}n_r + t_{\theta\theta}n_\theta = 0 \qquad (5.116)$$

where n_r and n_θ denote the components of the unit outward normal on the boundary. Since $n_r = -1$ and $n_\theta = 0$ on the boundary $r = a$, the above

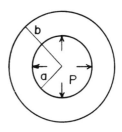

FIGURE 5.11.

conditions thus become simply

$$t_{rr} = -p, \qquad t_{r\theta} = 0 \tag{5.117}$$

at $r = a$.

In a similar way, we also require for the stress-free outside radius that

$$t_{rr} = 0, \qquad t_{r\theta} = 0 \tag{5.118}$$

at $r = b$.

In view of the above boundary conditions, we attempt a plane-strain, axisymmetric elasticity solution. The stress-function relation of equation (5.112) accordingly takes the form

$$\frac{d^4\phi}{dr^4} + \frac{2}{r} \frac{d^3\phi}{dr^3} - \frac{1}{r^2} \frac{d^2\phi}{dr^2} + \frac{1}{r^3} \frac{d\phi}{dr} = 0 \tag{5.119}$$

whose general solution is expressible as

$$\phi = A \log r + Br^2 \log r + Cr^2 + D \tag{5.120}$$

where A, B, C, and D denote constants.

The stress components are thus given from equation (5.113) as

$$t_{rr} = \frac{A}{r^2} + B(1 + 2 \log r) + 2C$$

$$t_{\theta\theta} = \frac{-A}{r^2} + B(3 + 2 \log r) + 2C \tag{5.121}$$

$$t_{r\theta} = 0$$

The above equations contain three unknown constants A, B, and C. Since the boundary conditions of equations (5.117) and (5.118) provide only two restrictions on these constants, the stress distribution is not uniquely determined by the stress conditions alone. This suggests that we need to examine also the displacement components before a unique solution can be obtained.

From the constitutive relations of equation (5.114) and the strain-displacement relations of Chapter 2, we have, for the case of *axially symmetric displacements*, the following relations:

$$e_{rr} = \frac{du_r}{dr} = \frac{1 + v}{E} [(1 - v)t_{rr} - vt_{\theta\theta}] \tag{5.122}$$

$$e_{\theta\theta} = \frac{u_r}{r} = \frac{1 + v}{E} [(1 - v)t_{\theta\theta} - vt_{rr}] \tag{5.123}$$

$$e_{r\theta} = \frac{du_\theta}{dr} - \frac{u_\theta}{r} = 0 \tag{5.124}$$

Substituting the previous expressions for t_{rr} and $t_{\theta\theta}$ into the first of these equations and integrating, we find on comparing with the second that the constant B must vanish and that u_r must be expressible as

$$u_r = \frac{1+\nu}{E}\left[-\frac{A}{r} + 2C(1-2\nu)r\right] \tag{5.125}$$

In addition, equation (5.124) gives for the displacement u_θ the relation

$$u_\theta = Fr \tag{5.126}$$

where F denotes a constant. This expression represents nothing more than a rigid rotation of the cylinder about its axis, so that, if its ends are assumed fixed against rotations of this kind, we must accordingly take $F = 0$ in this equation.

Returning to the stress relations of equation (5.121) and setting $B = 0$ in accordance with the above results, we may now establish, from the boundary condition of equations (5.117) and (5.118), the relations

$$A = -\frac{a^2 b^2}{b^2 - a^2}\, p, \qquad 2C = \frac{a^2}{b^2 - a^2}\, p$$

so that the stress and displacement components resulting from the internal pressure are finally determined as

$$t_{rr} = \frac{a^2 p}{b^2 - a^2}\left(1 - \frac{b^2}{r^2}\right)$$

$$t_{\theta\theta} = \frac{a^2 p}{b^2 - a^2}\left(1 + \frac{b^2}{r^2}\right) \tag{5.127}$$

$$t_{r\theta} = 0$$

and

$$u_r = \frac{1+\nu}{E}\,\frac{a^2 p}{b^2 - a^2}\left(1 + \frac{b^2}{r^2} - 2\nu\right)$$

$$u_\theta = 0 \tag{5.128}$$

It is interesting to note from the above solution that the stress component t_{rr} is always compressive and the component $t_{\theta\theta}$ always tensile regardless of the place considered in the cylinder. It is also interesting to note that the maximum value of the stress $t_{\theta\theta}$ is always greater than the internal pressure p and can never be made smaller than this pressure no matter how much the outer radius of the cylinder is increased.

5.9. Circular Hole in a Strained Plate

As an illustration of the use of the equations of plane stress in polar coordinates, we consider the important problem of stress concentrations arising from the presence of a circular hole in a thin plate under uniaxial loading as shown in Figure 5.12.

At the hole radius, we assume the boundary to be free from applied forces so that we require at $r = a$ that

$$t_{rr} = 0, \qquad t_{r\theta} = 0 \tag{5.129}$$

Also, at a large distance from the hole, we require the stress in the plate to approach that associated with the uniform axial loading $t_{xx} = S$. If we transform this stress component to axes oriented along the r and θ directions, we thus find for boundary conditions as $r \to \infty$ that

$$t_{rr} = \tfrac{1}{2}S(1 + \cos 2\theta)$$
$$t_{\theta\theta} = \tfrac{1}{2}S(1 - \cos 2\theta) \tag{5.130}$$
$$t_{r\theta} = \tfrac{1}{2}S \sin 2\theta$$

Using the above boundary conditions as a guide, we accordingly attempt a stress-function solution in the form

$$\phi = f(r) + g(r) \cos 2\theta \tag{5.131}$$

where $f(r)$ and $g(r)$ denote arbitrary functions of r only. On substituting into the governing stress-function relation of equation (5.112), we find that such solutions are, in fact, possible provided $f(r)$ and $g(r)$ satisfy

$$\left(\frac{d^2}{dr^2} + \frac{1}{r}\frac{d}{dr} \right)\left(\frac{d^2f}{dr^2} + \frac{1}{r}\frac{df}{dr} \right) = 0 \tag{5.132}$$

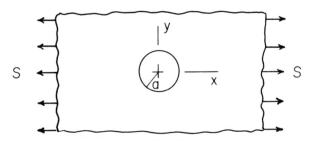

FIGURE 5.12.

and

$$\left(\frac{d^2}{dr^2} + \frac{1}{r}\frac{d}{dr} - \frac{4}{r^2}\right)\left(\frac{d^2g}{dr^2} + \frac{1}{r}\frac{dg}{dr} - \frac{4g}{r^2}\right) = 0 \tag{5.133}$$

The solutions of these equations may be written as

$$f = A \log r + Br^2 \log r + Cr^2 + D \tag{5.134}$$

and

$$g = Er^2 + Fr^4 + G/r^2 + H \tag{5.135}$$

where A, B, $\ldots$, H denote constants.

On combining these relations with equation (5.131) and calculating the stress components t_{rr}, $t_{\theta\theta}$, and $t_{r\theta}$ from equation (5.113), we find

$$t_{rr} = 2C + \frac{A}{r^2} - \left(2E + \frac{6G}{r^4} + \frac{4H}{r^2}\right)\cos 2\theta$$

$$t_{\theta\theta} = 2C - \frac{A}{r^2} + \left(2E + \frac{6G}{r^4}\right)\cos 2\theta \tag{5.136}$$

$$t_{r\theta} = \left(2E - \frac{6G}{r^4} - \frac{2H}{r^2}\right)\sin 2\theta$$

where the constants B and F in the above expressions have been set equal to zero in order to avoid large stresses for large r.

Finally, on using the boundary conditions of equations (5.129) and (5.130), we may evaluate the remaining five constants in the above expressions to find the stresses expressible as

$$t_{rr} = \frac{S}{2}\left(1 - \frac{a^2}{r^2}\right) + \frac{S}{2}\left(1 + \frac{3a^4}{r^4} - \frac{4a^2}{r^2}\right)\cos 2\theta$$

$$t_{\theta\theta} = \frac{S}{2}\left(1 + \frac{a^2}{r^2}\right) - \frac{S}{2}\left(1 + \frac{3a^4}{r^4}\right)\cos 2\theta \tag{5.137}$$

$$t_{r\theta} = -\frac{S}{2}\left(1 - \frac{3a^4}{r^4} + \frac{2a^2}{r^2}\right)\sin 2\theta$$

From these relations, it can be seen that the maximum stress $t_{\theta\theta}$ occurs at the ends of the hole diameter perpendicular to the direction of the uniaxial loading and, at these points, $t_{\theta\theta}$ is *three times the uniform axial stress*. This high stress concentration is obviously of great practical importance in the proper design of load-carrying members having holes drilled in them for one reason or another.

5.10. Strength-of-Material Formulations

In discussing the deformation of beams, plates, and shells under transverse normal loadings, it is frequently convenient to abandon some of the conditions imposed by exact elasticity theory and consider instead a simplified formulation. Treatments of this kind are known as *strength-of-material treatments* and are based for the most part on the observation that the bending associated with transverse normal loadings is governed primarily by the extensions and contractions of longitudinal filaments of the material and therefore that transverse normal and shearing strains may generally be neglected.

This situation is illustrated in Figure 5.13 for the case of bending of an elastic strip under compressive loading. From this figure, it can easily be seen that the upper filaments of the strip must be compressed and the lower ones extended as a direct consequence of the geometry.

The neglect of strains other than those associated with the extensions and contractions of the above filaments of material is, of course, equivalent to treating the actual elastic material as one which exhibits an infinite rigidity to transverse normal and shearing strains. For consistency, it is thus necessary that the constitutive relations describing the response of the material reflect this fact. Consider, in particular, the thin elastic element of material shown in Figure 5.14. Assuming for the moment that this material has different elastic properties in the three coordinate directions, we may use the results of the example of Section 4.6 for an orthotropic material to express the strains in terms of the stresses as

$$e_{xx} = S_{1111}t_{xx} + S_{1122}t_{yy} + S_{1133}t_{zz}$$
$$e_{yy} = S_{2211}t_{xx} + S_{2222}t_{yy} + S_{2233}t_{zz}$$
$$e_{zz} = S_{3311}t_{xx} + S_{3322}t_{yy} + S_{3333}t_{zz}$$
$$e_{xy} = S_{1212}t_{xy}$$
$$e_{yz} = S_{2323}t_{yz}$$
$$e_{xz} = S_{1313}t_{xz}$$

where S_{1111}, S_{1122}, etc., denote constants.

If we now require the material to be rigid in the z direction, we must obviously choose

$$S_{1133} = S_{2233} = S_{3311} = S_{3322} = S_{3333} = S_{2323} = S_{1313} = 0$$

In addition, for isotropic elastic response in the x–y plane, the material

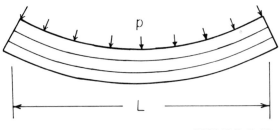

FIGURE 5.13.

symmetry arguments of Chapter 4 show that we must take

$$S_{1111} = S_{2222}, \qquad S_{1122} = S_{2211}, \qquad S_{1212} = S_{1111} - S_{1122}$$

Under these conditions, the above constitutive relations thus become expressible simply as

$$e_{xx} = S_{1111}t_{xx} + S_{1122}t_{yy}$$
$$e_{yy} = S_{1122}t_{xx} + S_{1111}t_{yy}$$
$$e_{xy} = (S_{1111} - S_{1122})t_{xy}$$

To relate the constants in these equations to known elastic properties of a material, we *assume* that a simple tension test on the above hypothetical material (with $t_{yy} = t_{xy} = 0$) would yield the same proportionality constants in the expressions

$$e_{xx} = S_{1111}t_{xx}, \qquad e_{yy} = \frac{S_{1122}}{S_{1111}} e_{xx}$$

as is found in testing a normal elastic material; that is,

$$S_{1111} = \frac{1}{E}, \qquad \frac{S_{1122}}{S_{1111}} = -\nu$$

where E and ν denote, of course, Young's modulus and Poisson's ratio.

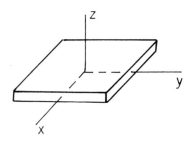

FIGURE 5.14.

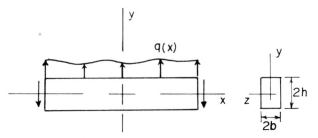

FIGURE 5.15.

Substituting these constants into the above equations, we thus have finally the governing constitutive relations for the hypothetical material expressible as

$$e_{xx} = \frac{1}{E}(t_{xx} - \nu t_{yy})$$

$$e_{yy} = \frac{1}{E}(t_{yy} - \nu t_{xx}) \qquad (5.138)$$

$$e_{xy} = \frac{1+\nu}{E} t_{xy}$$

It is worth noting that these equation are formally identical to those found by assuming a state of plane stress in the x–y plane. They were, of course, derived here without making such an assumption.

5.11. Bending and Extension of Beams

We consider the case of the static in-plane bending and extension of the initially straight beam of rectangular cross section shown in Figure 5.15. As indicated, we choose coordinates $X_1 = x$, $X_2 = y$ to denote initial position of particles in the beam and take the origin of these coordinates such that the x axis lies along the initial position of the centerline of the beam.

In accordance with earlier remarks, we assume that the beam material exhibits an infinite rigidity to transverse forces such that the strain components e_{yy} and e_{xy} always vanish regardless of the lateral loadings involved. In these circumstances, the displacement components $u_1 = u$, $u_2 = v$ for small deformations must then satisfy the relations

$$e_{yy} = \frac{\partial v}{\partial y} = 0$$

$$2e_{xy} = \frac{\partial u}{\partial y} + \frac{\partial v}{\partial x} = 0 \qquad (5.139)$$

The equilibrium equations may next be written in terms of the nonzero stresses t_{xx}, t_{yy}, and t_{xy} as

$$\frac{\partial t_{xx}}{\partial x} + \frac{\partial t_{xy}}{\partial y} = 0 \tag{5.140}$$

$$\frac{\partial t_{xy}}{\partial x} + \frac{\partial t_{yy}}{\partial y} = 0 \tag{5.141}$$

where body forces have been assumed absent.

If we count the number of unknowns appearing in the above equations, we find three components of stress and two components of displacement, giving a total of five unknowns in all. On the other hand, equations (5.139)–(5.141) provide only four equations connecting these quantities, so that we need one other relation before attempting a solution. This additional relation is, of course, supplied by a constitutive relation from equation (5.138) of the previous section. For the present problem where the y axis lies in the assumed rigid direction and where $t_{xz} = t_{yz} = t_{zz} = 0$, this relation takes the simple form

$$t_{xx} = E e_{xx} \tag{5.142}$$

where $e_{xx} = \partial u / \partial x$ denotes the axial strain in the beam.

Having established the above equations, it now remains to cast them in a convenient mathematical form for determining the displacements and stresses under arbitrary loading conditions. For this purpose, it is useful first to integrate the displacement relations of equations (5.139) and obtain displacement components for plane motion in the form

$$u = u_0(x) - y \frac{dv_0}{dx}, \qquad v = v_0(x) \tag{5.143}$$

where u_0 and v_0 denote the horizontal and vertical displacement components of the centerline of the beam. It is interesting to note that these components correspond to the condition that, on bending, plane sections of the beam remain plane and simply rotate about the centerline.

With these displacement relations, we may next integrate the equilibrium relations of equations (5.140) and (5.141) over the initial sectional area A of the beam. Using the boundary conditions

$$t_{yy} = \frac{q(x)}{2b}, \qquad t_{xy} = 0 \tag{5.144}$$

on $y = +h$ and

$$t_{yy} = 0, \qquad t_{xy} = 0 \qquad\qquad (5.145)$$

on $y = -h$, this integration yields the expressions

$$\frac{dN_{xx}}{dx} = 0 \qquad\qquad (5.146)$$

and

$$\frac{dN_{xy}}{dx} = -q(x) \qquad\qquad (5.147)$$

where $q(x)$ denotes the lateral loading per unit of initial length and where N_{xx} and N_{xy} are given by stress-resultant equations of the form

$$N_{xx} = \int_A t_{xx} \, dA \qquad\qquad (5.148)$$

and

$$N_{xy} = \int_A t_{xy} \, dA \qquad\qquad (5.149)$$

In addition, if we multiply equation (5.140) through by y and again integrate over the initial area A, we find the relation

$$\frac{dM_{xx}}{dx} = N_{xy} \qquad\qquad (5.150)$$

where M_{xx} is given by the resultant moment equation

$$M_{xx} = \int_A yt_{xx} \, dA \qquad\qquad (5.151)$$

and where, in performing the integration, use has again been made of the boundary conditions of equations (5.144) and (5.145).

If we use equations (5.142) and (5.143) with (5.151), we find further that the internal moment M_{xx} is expressible as

$$M_{xx} = E \frac{du_0}{dx} \int_A y \, dA - E \frac{d^2 v_0}{dx^2} \int_A y^2 \, dA \qquad\qquad (5.152)$$

However, since y is measured from the centroid of the cross section of the beam, we have

$$\int_A y \, dA = 0 \qquad\qquad (5.153)$$

and

$$I = \int_A y^2 \, dA = \tfrac{4}{3}bh^3 \tag{5.154}$$

so that the expression for M_{xx} becomes simply

$$M_{xx} = -EI \frac{d^2 v_0}{dx^2} \tag{5.155}$$

This relation is known as the *Bernoulli–Euler moment-curvature relation.*
A similar calculation also shows that the stress resultant N_{xx} is expressible as

$$N_{xx} = EA \frac{du_0}{dx} \tag{5.156}$$

On substituting this relation into equation (5.146), we thus have the governing equation for the centerline extension of the beam given as

$$\frac{d^2 u_0}{dx^2} = 0 \tag{5.157}$$

Finally, on eliminating the transverse shear force N_{xy} from equations (5.147) and (5.150) and using (5.155), we find the governing equation for the transverse deflection of the centerline of the beam expressible as

$$EI \frac{d^4 v_0}{dx^4} = q(x) \tag{5.158}$$

The boundary conditions to be used in connection with the above equation will, of course, depend on the manner in which the beam is assumed fixed. For purposes of illustration, we consider here three typical cases.

For the case of a *simply supported end* the vertical deflections of the centerline and the moments must vanish, so that, with equation (5.155), we have as end conditions at, say, $x = L$

$$v_0 = \frac{d^2 v_0}{dx^2} = 0 \tag{5.159}$$

In addition, depending on the longitudinal constraint, the axial displacements must also satisfy either

$$u_0 = 0 \tag{5.160a}$$

or

$$EA \frac{du_0}{dx} = N \qquad (5.160b)$$

at $x = L$, where N denotes an applied axial force, which may, of course, be set equal to zero for the case where end loadings are absent.

For the case of a *built-in end*, the vertical deflection and slope of the centerline of the beam must vanish, as well as the axial deflections, so that at $x = L$ we require

$$v_0 = \frac{dv_0}{dx} = u_0 = 0 \qquad (5.161)$$

Finally, for the case of a *free end*, the shear force N_{xy} and moment M_{xx} must vanish, as well as the axial force N_{xx}, so that at $x = L$ we require from equations (5.150), (5.155), and (5.156) that

$$\frac{d^2 v_0}{dx^2} = \frac{d^3 v_0}{dx^3} = \frac{du_0}{dx} = 0 \qquad (5.162)$$

After solving the above governing equations for the displacement components u_0 and v_0 using the appropriate boundary conditions, we can then find the stresses at points within the beam from the relations (5.140)–(5.143). In particular, with the displacement components known, the stress t_{xx} can be determined directly from equations (5.142) and (5.143) and this, in turn, used in the equilibrium relation of equation (5.140) to determine the shearing stress t_{xy}. Finally, with a knowledge of t_{xy}, the second equilibrium relation can then be integrated to yield the normal stress t_{yy}. With the above theory, we are thus able to determine the displacement and stresses in a beam subjected to arbitrary in-plane lateral and end loadings.

It is worth noting that equations (5.157) and (5.158) are also customarily applied to beams whose cross sections are symmetric about the y axis but whose sectional forms are not necessarily solid rectangles. In this case, the vertical coordinate y is measured from the centroid of the section and the quantity I is evaluated according to the integral of equation (5.154). Such treatments are, however, usually only approximate since stress boundary conditions on the lateral faces of the beam are generally not all satisfied.

Application to Uniform Loading.   As an example of the use of the above equations, we consider the problem of uniform loading of a simply supported beam. This problem, which was considered in Section 5.6 using plane-stress elasticity theory, is illustrated in Figure 5.16.

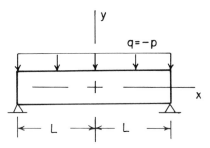

FIGURE 5.16.

The boundary conditions at $x = \pm L$ can be seen to be identical to those given by equations (5.159) and (5.160) with longitudinal end displacements equal to zero, namely

$$v_0 = \frac{d^2v_0}{dx^2} = u_0 = 0 \qquad \text{at} \quad x = \pm L \tag{5.163}$$

Using the last of these boundary conditions in connection with equations (5.156) and (5.157), it can immediately be seen that the stress resultant N_{xx} must vanish everywhere along the beam. In addition, the remaining transverse deflection relation given by equation (5.158) is expressible for this case as

$$\frac{d^4v_0}{dx^4} = -\frac{p}{EI} \tag{5.164}$$

where p denotes the uniform transverse compressive loading as shown in Figure 5.16. Integrating this equation four times and using the boundary conditions of equation (5.163), we find

$$v_0 = \frac{pL^4}{2EI}\left(\frac{1}{2}\frac{x^2}{L^2} - \frac{1}{12}\frac{x^4}{L^4} - \frac{5}{12}\right) \tag{5.165}$$

On substituting this result into equation (5.143) and using (5.142), we next find that the stress component t_{xx} associated with the bending can be expressed as

$$t_{xx} = \frac{py}{2I}(x^2 - L^2) \tag{5.166}$$

Substituting this relation into the equilibrium relation given by equation

(5.140) and integrating, we find further that the shear stress component t_{xy} is determined by

$$t_{xy} = -\frac{px}{2I}(y^2 - h^2) \qquad (5.167)$$

where $2h$ denotes the height of the beam and where use has been made of the shear stress boundary conditions of equations (5.144) and (5.145). Similarly substituting this relation into the equilibrium equation (5.141) and integrating, we find with the help of the boundary conditions of equations (5.144) and (5.145) that the normal stress component t_{yy} is given by

$$t_{yy} = \frac{p}{2I}\left(\frac{y^3}{3} - h^2 y - \frac{2h^3}{3}\right) \qquad (5.168)$$

Having the above solutions before us, we may now compare them with the corresponding results established in Section 5.6 using plane-stress elasticity theory. From that solution, we have, in particular, the stress t_{xx} given by

$$t_{xx} = \frac{p}{2I}(x^2 - L^2)y - \frac{p}{2I}\left(\frac{2}{3}y^3 - \frac{2}{5}h^2 y\right) \qquad (5.169)$$

This expression is not identical to that determined by the beam theory, but comparison with equation (5.166) shows that the two differ only by the last term in equation (5.169), whose magnitude is of the order of h^2/L^2 when compared with the maximum bending stress. When the beam is long in comparison with its height, this term will accordingly be small and we may therefore regard the stress t_{xx} given by the beam theory as an accurate estimate of the actual stress.

Also, if we compare the remaining stresses determined by beam theory with those given by the elasticity solution, we find that the stress components t_{xy} and t_{yy} determined by the beam theory are identical to those derived from elasticity theory.

From the above results, we are thus led to expect that the restriction to a hypothetical elastic material exhibiting infinite rigidity to transverse strains will yield an accurate description of the bending stresses and deflections of a beam, or an elastic strip taken from a plate or cylindrical shell, provided its length is large in comparison with its transverse dimensions, or, more generally, provided that the characteristic length of the deflection distribution is large in comparison with the transverse dimensions.

5.12. Bending and Extension of Thin Rectangular Plates

The case of a thin rectangular plate subjected to lateral loading (Figure 5.17) may be treated in a manner analogous to that employed for beams in the previous section. Thus, we assume the material to exhibit an infinite rigidity to transverse normal and shearing strains such that

$$e_{zz} = \frac{\partial w}{\partial z} = 0$$

$$2e_{xz} = \frac{\partial u}{\partial z} + \frac{\partial w}{\partial x} = 0 \tag{5.170}$$

$$2e_{yz} = \frac{\partial v}{\partial z} + \frac{\partial w}{\partial y} = 0$$

where $u = u_1$, $v = u_2$, $w = u_3$ denote displacement components.

The equilibrium equations for this problem may likewise be written as

$$\frac{\partial t_{xx}}{\partial x} + \frac{\partial t_{xy}}{\partial y} + \frac{\partial t_{xz}}{\partial z} = 0$$

$$\frac{\partial t_{xy}}{\partial x} + \frac{\partial t_{yy}}{\partial y} + \frac{\partial t_{yz}}{\partial z} = 0 \tag{5.171}$$

$$\frac{\partial t_{xz}}{\partial x} + \frac{\partial t_{yz}}{\partial y} + \frac{\partial t_{zz}}{\partial z} = 0$$

where body forces have been assumed absent and where the angular momentum relations $t_{xy} = t_{yx}$, $t_{xz} = t_{zx}$, and $t_{yz} = t_{zy}$ have been used.

Finally, the constitutive relations are given by equation (5.138) of Section 5.13. Using the strain-displacement relations $e_{xx} = \partial u/\partial x$, etc.,

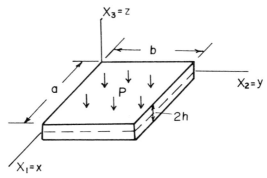

FIGURE 5.17.

these may be written as

$$t_{xx} = \frac{E}{1 - v^2} \left(\frac{\partial u}{\partial x} + v \frac{\partial v}{\partial y} \right)$$

$$t_{yy} = \frac{E}{1 - v^2} \left(\frac{\partial v}{\partial y} + v \frac{\partial u}{\partial x} \right) \tag{5.172}$$

$$t_{xy} = \frac{E}{2(1 + v)} \left(\frac{\partial u}{\partial y} + \frac{\partial v}{\partial x} \right)$$

The set of equations (5.170)–(5.172) provide nine relations for determining the six unknown components of stress and the three unknown components of displacement. In order to cast them in a convenient mathematical form, we first integrate equations (5.170) to get

$$u = u_0 - z \frac{\partial w_0}{\partial x}$$

$$v = v_0 - z \frac{\partial w_0}{\partial y} \tag{5.173}$$

$$w = w_0$$

where u_0, v_0, and w_0 denote displacements of the midplane of the plate where $z = 0$.

Next we note that the boundary conditions require that on the plate surface $z = +h$

$$t_{xz} = t_{yz} = 0, \qquad t_{zz} = -P(x, y) \tag{5.174}$$

and that on the surface $z = -h$

$$t_{xz} = t_{yz} = t_{zz} = 0 \tag{5.175}$$

Using these conditions, we can integrate equations (5.171) over the thickness of the plate to obtain

$$\frac{\partial N_{xx}}{\partial x} + \frac{\partial N_{xy}}{\partial y} = 0$$

$$\frac{\partial N_{xy}}{\partial x} + \frac{\partial N_{yy}}{\partial y} = 0 \tag{5.176}$$

$$\frac{\partial N_{xz}}{\partial x} + \frac{\partial N_{yz}}{\partial y} = P$$

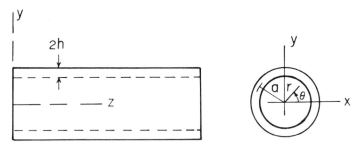

FIGURE 5.19.

constant, the coefficients A_{mn} may be determined from equation (5.194) simply as

$$A_{mn} = \frac{16P_0}{\pi^2 mn} \qquad (5.196)$$

so that

$$C_{mn} = -\frac{16P_0}{D\pi^6 mn}\left(\frac{a^2 b^2}{m^2 b^2 + n^2 a^2}\right)^2 \qquad (5.197)$$

With this result, the deflection at any point x, y in the midplane may then easily be worked out using the series solution of equation (5.192).

Once the deflection equation has been determined for any specified loading, the stresses may then be determined using the constitutive and equilibrium equations. In particular, with $w_0(x, y)$ known, equations (5.173) provide the displacement components at all points in the plate and these may thus be used with the constitutive relations of equation (5.172) to determine the stresses t_{xx}, t_{yy}, and t_{xy}. Finally, using these in the stress equilibrium relations of equations (5.171), we may find by integration the remaining stresses t_{xz}, t_{yz}, and t_{zz}.

5.13. Axisymmetric Bending and Extension of Thin Cylindrical Shells

We consider finally the case of a thin, circular cylindrical shell subjected to axisymmetric loading. The initial mid-surface radius of the shell is denoted by a and its thickness by $2h$ as shown in Figure 5.19.

We denote the initial position of particles in the shell by the rectangular coordinates $X_1 = x$, $X_2 = y$, $X_3 = z$ and introduce cylindrical polar coordinates r, θ, z such that

$$x = r\cos\theta, \qquad y = r\sin\theta, \qquad z = z$$

From the results given in Chapter 2, we may write the strain-displacement relations in polar coordinates for the case of axisymmetric deformations as follows:

$$e_{rr} = \frac{\partial u}{\partial r}, \qquad e_{\theta\theta} = \frac{u}{r}, \qquad e_{zz} = \frac{\partial w}{\partial z}$$

$$2e_{rz} = \frac{\partial u}{\partial z} + \frac{\partial w}{\partial r}, \qquad e_{\theta z} = e_{r\theta} = 0$$

(5.198)

where u and w have been used to denote the radial and axial components of displacement.

In order to describe the radial displacement of the shell resulting from bending under lateral loadings, we assume the shell material to exhibit an infinite rigidity to the transverse strains e_{rr}, $e_{r\theta}$, and e_{rz}. From equation (5.198), it thus follows that

$$e_{rr} = \frac{\partial u}{\partial r} = 0$$

$$2e_{rz} = \frac{\partial u}{\partial z} + \frac{\partial w}{\partial r} = 0$$

(5.199)

In addition to these assumed relations, we also have equations expressing the equilibrium of the shell. From the results of Chapter 3, these are known to be expressible for axisymmetric loadings as

$$\frac{\partial t_{rr}}{\partial r} + \frac{\partial t_{rz}}{\partial z} + \frac{t_{rr} - t_{\theta\theta}}{r} = 0$$

$$\frac{\partial t_{rz}}{\partial r} + \frac{\partial t_{zz}}{\partial z} + \frac{t_{rz}}{r} = 0$$

(5.200)

where t_{rr}, t_{rz}, and $t_{\theta\theta}$ denote polar stress components.

Finally, we have the constitutive relations of equation (5.138), which here take the form

$$t_{\theta\theta} = \frac{E}{1 - v^2} (e_{\theta\theta} + v e_{zz})$$

$$t_{zz} = \frac{E}{1 - v^2} (e_{zz} + v e_{\theta\theta})$$

(5.201)

where E and v again denote Young's modulus and Poisson's ratio.

Having the above six equations, we now proceed to develop a method of solution for various axisymmetric loading conditions. If we integrate equations (5.199), we find immediately that the displacement components must satisfy

$$u = u_0(z)$$

$$w = w_0(z) - (r - a)\frac{\partial u_0}{\partial z}$$

(5.202)

where u_0 and w_0 denote the displacement components of the mid-surface of the shell.

In addition to these relations, we may also introduce boundary conditions such that on $r = a - h$

$$t_{rr} = -P(z), \qquad t_{rz} = 0$$

(5.203)

and on $r = a + h$,

$$t_{rr} = 0, \qquad t_{rz} = 0$$

(5.204)

On multiplying equations (5.200) through by $r\,dr$ and integrating between the limits $a - h$ and $a + h$, we find with the help of these boundary conditions and equations (5.202) that

$$\frac{dN_{rz}}{dz} - \frac{N_{\theta\theta}}{a} + P = 0$$

(5.205)

and

$$\frac{dN_{zz}}{dz} = 0$$

(5.206)

where the stress resultants N_{rz} and N_{zz} per unit of initial circumferential length and the stress resultant $N_{\theta\theta}$ per unit of initial length are given by

$$N_{rz} = \frac{1}{a}\int_{a-h}^{a+h} t_{rz}r\,dr$$

(5.207)

$$N_{zz} = \frac{1}{a}\int_{a-h}^{a+h} t_{zz}r\,dr$$

(5.208)

$$N_{\theta\theta} = \int_{a-h}^{a+h} t_{\theta\theta}\,dr$$

(5.209)

In carrying out the integrations involved in obtaining equations (5.205) and (5.206), use has also been made of the assumption of a thin shell such that h/a is small in comparison with unity.

On multiplying the second of equations (5.200) through by $(r - a)r\,dr$ and integrating again over the thickness of the shell, we find further with the help of the boundary conditions of equations (5.203) and (5.204) and equation (5.202) that

$$\frac{dM_{zz}}{dz} = N_{rz} \qquad (5.210)$$

where the internal moment M_{zz} per unit of initial circumferential length is given by

$$M_{zz} = \frac{1}{a} \int_{a-h}^{a+h} (r - a)t_{zz}r\,dr \qquad (5.211)$$

and where, in carrying out the integration associated with equation (5.210), use has again been made of the fact that the ratio h/a is small.

The above stress resultants and moment may readily be evaluated using equations (5.201) and (5.202). We find, in particular, that

$$N_{\theta\theta} = \frac{Eh}{1 - \nu^2} \left(\frac{u_0}{a} + \nu\,\frac{\partial w_0}{\partial z} \right) \qquad (5.212)$$

$$N_{zz} = \frac{2Eh}{1 - \nu^2} \left(\frac{\partial w_0}{\partial z} + \nu\,\frac{u_0}{a} \right) \qquad (5.213)$$

$$M_{zz} = -D \left(\frac{\partial^2 u_0}{\partial z^2} - \frac{1}{a}\,\frac{\partial w}{\partial z} \right) \qquad (5.214)$$

where

$$D = \frac{2}{3}\,\frac{Eh^3}{1 - \nu^2}$$

Substituting the above expression for N_{zz} into the equilibrium relation of equation (5.206), we thus have

$$\frac{d^2 w_0}{dz^2} + \frac{\nu}{a}\,\frac{du_0}{dz} = 0 \qquad (5.215)$$

which represents the governing equation for the axial displacement w_0 of the mid-surface of the shell.

Similarly, on eliminating the transverse shear force N_{rz} between equations (5.205) and (5.210) and using the above moment-displacement relation, we find

$$D\left(\frac{d^4 u_0}{dz^4} - \frac{1}{a}\,\frac{d^3 w_0}{dz^3} \right) + \frac{2Eh}{1 - \nu^2} \left(\frac{u_0}{a^2} + \frac{\nu}{a}\,\frac{dw_0}{dz} \right) = P \qquad (5.216)$$

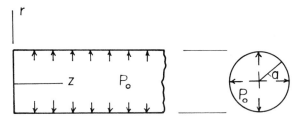

FIGURE 5.20.

which represents the governing equation for radial displacements of the shell mid-surface.

The boundary conditions to be employed with the above equations are similar to those already discussed in connection with beam theory. We mention here only the simple case of a built-in edge constrained against displacement and rotation where we require at, say, $z = l$, that

$$u_0 = \frac{du_0}{dz} = w_0 = 0 \qquad (5.217)$$

Having solutions for the displacements, we may, of course, return to the equilibrium and constitutive relations to determine the stresses existing in the shell in the same fashion as discussed earlier for beams.

Application to Uniform Pressure Loading. To illustrate the use of the above equations, we consider the problem of a long cylindrical tube subjected to a uniform internal pressure loading P_0 (Figure 5.20).

The governing equations for the axial and radial deflections of the mid-surface of the shell are determined from equations (5.125) and (5.216) as

$$\frac{d^2 w_0}{dz^2} + \frac{v}{a} \frac{du_0}{dz} = 0 \qquad (5.218)$$

and

$$D\left(\frac{d^4 u_0}{dz^4} - \frac{1}{a} \frac{d^3 w_0}{dz^3} \right) + \frac{2Eh}{1 - v^2} \left(\frac{u_0}{a^2} + \frac{v}{a} \frac{dw_0}{dz} \right) = P_0 \qquad (5.219)$$

The boundary conditions on the shell at $z = 0$ are also taken to be those given by equation (5.217), namely

$$u_0 = \frac{du_0}{dz} = w_0 = 0 \qquad (5.220)$$

From equation (5.218), we have by differentiation that

$$\frac{d^3 w_0}{dz^3} = -\frac{v}{a}\frac{d^2 u_0}{dz^2} \tag{5.221}$$

and by integration that

$$\frac{dw_0}{dz} = -v\frac{u_0}{a} \tag{5.222}$$

where, in this last equation, the constant of integration has been set equal to zero corresponding to the case of zero applied axial force N_{xx}. Hence, substituting these results into equation (5.219), we find

$$D\frac{d^4 u_0}{dz^4} + \frac{vD}{a^2}\frac{d^2 u_0}{dz^2} + \frac{2Eh}{a^2}u_0 = P_0 \tag{5.223}$$

which provides the governing equation for the radial deflection u_0.

The general solution to equation (5.233), valid for $h/a \ll 1$, may be written as

$$u_0 = e^{-\alpha z}(C_1 \cos \alpha z + C_2 \sin \alpha z)$$
$$+ e^{\alpha z}(C_3 \cos \alpha z + C_4 \sin \alpha z) + \frac{P_0 a^2}{2Eh} \tag{5.224}$$

where C_1, C_2, C_3, and C_4 denote constants and where α is given by

$$\alpha = \left(\frac{Eh}{2a^2 D}\right)^{1/4}$$

If we assume the shell to be of infinite length, we must obviously choose $C_3 = C_4 = 0$ in order to have bounded displacements, and this solution then reduces simply to

$$u_0 = e^{-\alpha z}(C_1 \cos \alpha z + C_2 \sin \alpha z) + \frac{P_0 a^2}{2Eh} \tag{5.225}$$

The constants C_1 and C_2 are determined from the boundary conditions of equation (5.220) as

$$C_1 = C_2 = -\frac{P_0 a^2}{2Eh} \tag{5.226}$$

so that the deflection is thus expressible as

$$u_0 = -\frac{P_0 a^2}{2Eh}[e^{-\alpha z}(\cos \alpha z + \sin \alpha z) - 1] \tag{5.227}$$

It will be seen that as the quantity αz increases, the radial deflection u_0 rapidly approaches the uniform deflection given by

$$u_0 = \frac{P_0 a^2}{2Eh} \tag{5.228}$$

This indicates that the variation in deflection produced by the end constraint is of local character only and that the above solution can therefore be used also for the case of finite-length tubes provided their lengths are such as to allow neglect of the effects of one end constraint on the other.

Having the above solution for the deflection, we may, of course, determine the stresses in the shell using equations (5.200)–(5.202). In particular, using equations (5.222) and (5.227), we may calculate the stresses t_{rr} and $t_{\theta\theta}$ directly from the displacement equations (5.202) and the constitutive equations (5.201). With these, we may then finally determine the remaining nonzero stresses t_{rz} and t_{zz} by direct integration of the stress-equilibrium equations as determined from equation (5.200).

Selected Reading

Timoshenko, S., and J. N. Goodier, *Theory of Elasticity*. McGraw-Hill Book Co., New York, 1951. An excellent book on the application of linear elasticity to isotropic solids.

Long, R. R., *Mechanics of Solids and Fluids*. Prentice-Hall, Englewood Cliffs, New Jersey, 1961. Chapter 6 discusses problems in linear elasticity.

Kolsky, H., *Stress Waves in Solids*. Clarendon Press, Oxford, England, 1953. A readable treatment of elastic wave propagation in solids.

Fung, Y. C., *Foundations of Solid Mechanics*. Prentice-Hall, Englewood Cliffs, New Jersey, 1965. Elasticity solutions are discussed in Chapters 7–9.

Timoshenko, S., and S. Woinowsky-Krieger, *Theory of Plates and Shells*. McGraw-Hill Book Co., New York, 1959. Approximate strength-of-material equations governing the bending of plates and shells are derived and applied to a number of technically important problems.

Koiter, W. T., and J. G. Simmonds, "Foundations of Shell Theory," Delft University of Technology, the Netherlands, 1972. Advanced discussion of shell theory presented at the 13th International Congress of Theoretical and Applied Mechanics, Moscow, 1972.

Exercises

1. Show that the warping function

$$\psi = \frac{b^2 - a^2}{a^2 + b^2} \, xy$$

solves the problem of the twisting of a solid bar of elliptic cross section defined by

$$\frac{x^2}{a^2} + \frac{y^2}{b^2} = 1$$

Show also that the moment-twist relation is

$$M = \frac{\mu\alpha\pi a^3 b^3}{a^2 + b^2}$$

2. Introduce the stress function ϕ such that

$$t_{xz} = \frac{\partial\phi}{\partial y}, \qquad t_{yz} = -\frac{\partial\phi}{\partial x}$$

and show that the twisting of prismatic bars can be discussed using the equation

$$\frac{\partial^2\phi}{\partial x^2} + \frac{\partial^2\phi}{\partial y^2} = -2\mu\alpha$$

and the boundary condition

$$\phi = c = \text{constant}$$

This is Prandtl's formulation.

3. Consider a rectangular beam subjected to in-plane moments M applied at its ends. Show that the equations of elasticity can all be satisfied by assuming only the axial stress nonzero and of the form

$$t_{xx} = ky$$

where k denotes a constant to be determined and y denotes vertical distance from the horizontal centerline of the beam.

4. Determine the strain components associated with the stress distribution of Exercise 3. Integrate these, using the strain-displacement relations, to determine the displacement components. Assume for boundary conditions that the beam is fixed against translation and rotation at $y = 0$ on the left-hand end.

5. Show that the stress in a long cylinder having internal pressure loading and stress-free ends can be analyzed using the plane strain solution of equation (5.127) and an added axial stress.

6. Using an analysis similar to that in Section 5.8, determine the stresses and displacements in a thick-walled cylinder subjected to external pressure P_b.

7. Using the *principle of superposition* (Exercise 8 of Chapter 4), add the results of Exercise 6 and Section 5.8 to obtain the general solution for a cylinder subjected to both internal and external pressure.

8. Consider a beam of length L subjected to a concentrated downward load F applied a distance $x = a$ from the left end. By examining the geometry and equilibrium of a small element of the beam having sides $a - \Delta x$ and $a + \Delta x$ from the left end, show that the following conditions must apply at $x = a$:

$$v_a = v_b, \qquad v_a' = v_b', \qquad v_a'' = v_b'', \qquad v_a''' - v_b''' = -F/EI$$

where v_a and v_b denote general solutions (each containing four constants) of equation (5.158) for the regions $a < x < L$ and $0 < x < a$, respectively, and primes denote derivatives with respect to x.

9. Use the results of Exercise 8 to solve for the deflection of a simply supported beam having a concentrated downward force applied at $x = \alpha L$, where α denotes a constant such that $0 < \alpha < 1$.

PART III

THERMAL ELASTICITY

The theory of thermal elasticity brings together the separate theories of elasticity and thermodynamics in an effort to allow determination of the internal stresses and motions of an elastic solid subjected to general mechanical and thermal loadings.

The inclusion of thermal effects into the theory of linear elasticity is attributable chiefly to the work of *J. M. C. Duhamel* in France in 1838 and *Franz Neumann* in Germany in 1841. It was *Duhamel* who first noted that the elements in a solid body suffering temperature changes cannot expand freely and that thermal stresses must necessarily ensue. In studying these stresses, he developed linear equations of thermoelasticity in much the form as we know them today. *Neumann*, working independently, developed similar equations.

6

Theory of Thermal Elasticity

In the previous discussion of elastic deformations, the possibility of effects arising from temperature changes in the solid was completely ignored. From elementary physics, it is, however, known that changes in temperature in an unconstrained solid generally produce changes in its linear dimensions. These changes give rise, in turn, to corresponding strains and when such strains are prevented by boundary supports, the temperature changes must then induce stresses within the material. Since our earlier discussions did not deal with such strains or stresses, we were thus implicitly assuming the case of constant-temperature, or isothermal, deformations.

In order to be able to discuss the case of thermally induced stresses and strains in an elastic solid, we consider the theory of thermal elasticity in the present chapter, regarding it primarily as a supplement to the elasticity theory already discussed. We begin by considering the restrictions placed on the thermal effects by the first and second laws of thermodynamics.

6.1. First Law of Thermodynamics

The first law of thermodynamics introduces the concepts of *energy* and *work* and requires simply that the rate of increase of the total energy of a continuum mass be balanced by the rate at which work is done on it by applied forces and by the net rate at which thermal energy (heat) is supplied to it. To express this law in mathematical form, we consider an arbitrarily chosen portion of a continuum mass as shown in Figure 6.1.

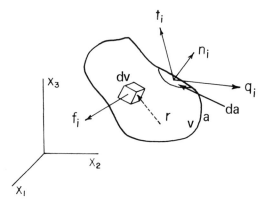

FIGURE 6.1.

For this continuum mass, the rate of work done by the surface and body forces is defined by the sum

$$\oint_a t_i v_i \, da + \int_v f_i v_i \, dv \tag{6.1}$$

where t_i denotes the Cauchy stress vector, f_i denotes the body force vector, and v_i denotes the velocity of the material at position x_i.

Similarly, the rate of increase of thermal energy in the continuum mass is defined as

$$-\oint_a q_i n_i \, da + \int_v \varrho r \, dv \tag{6.2}$$

where q_k denotes the heat flux (or flow) per unit area out of the body, n_i denotes the outward normal, ϱ denotes the mass density, and r denotes the heat supply rate per unit mass.

Finally, the rate of increase of the total energy of the mass is defined as

$$\frac{d}{dt} \int_v \frac{1}{2} \varrho v_i v_i \, dv + \frac{d}{dt} \int_v \varrho U \, dv \tag{6.3}$$

where the term $\frac{1}{2} \varrho v_i v_i$ denotes the kinetic energy per unit volume and U denotes the internal energy per unit mass.

Collecting together the above expressions, we thus have the *first law of thermodynamics* given as

$$\frac{d}{dt} \int_v \varrho \left(\frac{1}{2} v_i v_i + U \right) dv = \int_v (f_i v_i + \varrho r) \, dv + \oint_a (t_i v_i - q_i n_i) \, da \tag{6.4}$$

Since the above equation applies to any arbitrary choice of continuum mass, we may express it in local (differential) form following the same procedures as employed in Chapter 3 for the integral statements of mass conservation and momentum balance. With the help of the continuity and momentum equations, we find after some reduction that

$$\varrho \frac{dU}{dt} = \varrho r - \frac{\partial q_i}{\partial x_i} + t_{ji} \frac{\partial v_i}{\partial x_j} \tag{6.5}$$

which, accordingly, represents the *local form of the first law of thermodynamics*.

6.2. Second Law of Thermodynamics

The second law of thermodynamics introduces the concept of *entropy* and requires simply that the rate of increase of the total entropy of a continuum mass must always be equal to or greater than the ratio of the net rate of added heat to the absolute temperature. For the continuum mass shown in Figure 6.1, the *second law of thermodynamics* thus takes the form

$$\frac{d}{dt} \int_v \varrho S \, dv \geq \int_v \frac{\varrho r}{T} \, dv - \oint_a \frac{q_i n_i}{T} \, da \tag{6.6}$$

where S denotes the entropy per unit mass and where T denotes the absolute temperature. This relation is referred to as the *Clausius–Duhem inequality*. The equality applies only for so-called reversible processes.

In the same manner as for the first law, this relation can also be reduced to

$$\varrho T \frac{dS}{dt} - \varrho r + \frac{\partial q_i}{\partial x_i} - \frac{q_i}{T} \frac{\partial T}{\partial x_i} \geq 0 \tag{6.7}$$

which thus represents the *local form of the second law of thermodynamics*.

It is generally convenient to introduce into equation (6.7) the so-called Helmholtz function A defined by

$$A = U - TS \tag{6.8}$$

so that, on using the first law as expressed by equation (6.5), we may write an equivalent relation for the local form of the second law as

$$t_{ki} \frac{\partial v_i}{\partial x_k} - \varrho \left(S \frac{dT}{dt} + \frac{dA}{dt} \right) - \frac{q_i}{T} \frac{\partial T}{\partial x_i} \geq 0 \tag{6.9}$$

6.3. Definition of a Thermoelastic Solid

A *thermoelastic solid* is defined as one for which at each material point the following constitutive relations are taken to apply:

$$A = A\left(T, \frac{\partial x_i}{\partial X_j}, \frac{\partial T}{\partial X_j}\right) \tag{6.10}$$

$$S = S\left(T, \frac{\partial x_i}{\partial X_j}, \frac{\partial T}{\partial X_j}\right) \tag{6.11}$$

$$q_i = q_i\left(T, \frac{\partial x_i}{\partial X_j}, \frac{\partial T}{\partial X_j}\right) \tag{6.12}$$

$$t_{ij} = t_{ij}\left(T, \frac{\partial x_i}{\partial X_j}, \frac{\partial T}{\partial X_j}\right) \tag{6.13}$$

where X_i denotes the initial position of particles in the stress-free reference configuration that occupy position x_i at time t.

6.4. Restrictions Placed on Constitutive Relations by the Second Law of Thermodynamics

We now examine the restrictions placed on the general form of the above constitutive relations by the second law of thermodynamics. Introducing the notation

$$F_{ij} = \frac{\partial x_i}{\partial X_j}, \qquad G_j = \frac{\partial T}{\partial X_j} \tag{6.14}$$

and substituting the above constitutive relations into equation (6.9), we find

$$t_{ki}\frac{\partial v_i}{\partial x_k} - \varrho\left(S + \frac{\partial A}{\partial T}\right)\frac{dT}{dt} - \varrho\frac{\partial A}{\partial F_{ij}}\frac{dF_{ij}}{dt}$$
$$- \varrho\frac{\partial A}{\partial G_i}\frac{dG_i}{dt} - \frac{q_i}{T}\frac{\partial T}{\partial x_i} \geq 0 \tag{6.15}$$

If we fix all quantities appearing in the constitutive relations, we see from the above relation that we are still free to choose the magnitudes of dT/dt and dG_i/dt arbitrarily. Hence, in order to keep from violating the inequality by inappropriate choices of these quantities, we must thus take

$$S + \frac{\partial A}{\partial T} = \frac{\partial A}{\partial G_i} = 0 \tag{6.16}$$

From these relations, we therefore have that

$$S = -\frac{\partial A}{\partial T} \tag{6.17}$$

and that A and S are both independent of the temperature gradient G_i. Since the term dF_{ij}/dt in equation (6.15) may be written as

$$\frac{dF_{ij}}{dt} = \frac{\partial v_i}{\partial X_j} = \frac{\partial v_i}{\partial x_k} \frac{\partial x_k}{\partial X_j} \tag{6.18}$$

the part of this equation remaining after we have used equation (6.16) is expressible as

$$\left(t_{ki} - \varrho F_{kj} \frac{\partial A}{\partial F_{ij}}\right) \frac{\partial v_i}{\partial x_k} - \frac{q_i}{T} \frac{\partial T}{\partial x_i} \geq 0 \tag{6.19}$$

Now, since $\partial v_i/\partial x_k$ is not fixed by the variables in the constitutive relations, it may also be chosen arbitrarily and, hence, we must thus take

$$t_{ki} = \varrho F_{kj} \frac{\partial A}{\partial F_{ij}} \tag{6.20}$$

in order to prevent violation of the inequality by inappropriate choice of $\partial v_i/\partial x_k$. From this equation and the fact that A is independent of the temperature gradients, we thus see that the stress must also be independent of these gradients.

Finally, on taking equation (6.20) into account, the second law reduces simply to

$$q_i \frac{\partial T}{\partial x_i} \leq 0 \tag{6.21}$$

the absolute temperature T being always positive.

Collecting together the above results, we therefore have the constitutive relations for a thermoelastic solid expressible as

$$\begin{aligned} A &= A(T, F_{ij}) \\ S &= -\frac{\partial A}{\partial T} \\ q_i &= q_i(T, F_{ij}, G_j) \\ t_{ki} &= \varrho F_{kj} \frac{\partial A}{\partial F_{ij}} \end{aligned} \tag{6.22}$$

6.5. Restrictions Placed on Constitutive Relations by Principle of Material Indifference

We next consider restrictions placed on the above constitutive relations by the principle of material indifference as discussed in Chapter 4. For the Helmholtz function A given in equation (6.22), we require, in particular, that if

$$A = A(T, F_{ij}) \tag{6.23}$$

holds for the motion

$$x_i = x_i(X_j, t) \tag{6.24}$$

then

$$A = A(T, F'_{ij}) \tag{6.25}$$

must also hold for the motion

$$x'_i = Q_{ij}x_j + U_i \tag{6.26}$$

where Q_{ij} and U_i describe arbitrary time-dependent rigid-body rotations and translations of the material. Thus we require that

$$A(T, F_{ij}) = A(T, Q_{ik}F_{kj}) \tag{6.27}$$

Following the procedures of Chapter 4, we choose the rotation Q_{ij} to be equal and opposite to the rotation R_{ij} associated with the motion described by equation (6.24), namely

$$Q_{ij} = R_{ji} \tag{6.28}$$

so that equation (6.27) thus becomes expressible as

$$A(T, F_{ij}) = A(T, R_{ki}F_{kj}) \tag{6.29}$$

Now, from Chapter 2 we have that the deformation gradient F_{kj} can be expressed in terms of the rotation R_{kp} and a stretch U_{pj} as

$$F_{kj} = R_{kp}U_{pj} \tag{6.30}$$

Hence, multiplying through by R_{ki} and using the orthogonality condition

$$R_{ki}R_{kp} = \delta_{ip} \tag{6.31}$$

we have

$$R_{ki}F_{kj} = U_{ij} \tag{6.32}$$

so that equation (6.29) becomes simply

$$A = A(T, U_{ij}) \tag{6.33}$$

Alternatively, since U_{ij} is a function of C_{ij}, where

$$C_{ij} = F_{mi}F_{mj} = \frac{\partial x_m}{\partial X_i} \frac{\partial x_m}{\partial X_j} \tag{6.34}$$

equation (6.34) may also be written as

$$A = A(T, C_{ij}) \tag{6.35}$$

Similarly, we may also show using these same procedures that the principle of material indifference requires the heat flux vector q_k to take the form

$$q_i = R_{ik}q_k(T, G_j, C_{jk}) \tag{6.36}$$

where R_{ik} denotes the rotation associated with the deformation gradients F_{ik}.

On substituting the above expression for the Helmholtz function A into the second and fourth relations of equation (6.22), we find the constitutive relations for a thermoelastic solid expressible as

$$
\begin{aligned}
A &= A(T, C_{jk}) \\
S &= -\frac{\partial A}{\partial T} \\
q_i &= R_{ik}q_k(T, G_j, C_{jk}) \\
t_{ki} &= \varrho F_{kj}F_{il}\left(\frac{\partial A}{\partial C_{jl}} + \frac{\partial A}{\partial C_{lj}}\right)
\end{aligned}
\tag{6.37}
$$

6.6. Restriction to Small Deformations and Small Temperature Changes

For small deformations and small temperature changes, the above constitutive relations may be linearized following the same general procedure as outlined in Chapter 4. Under these circumstances, we may, in particular, choose x_i and T to be given by

$$x_i = X_i + u_i \tag{6.38}$$

$$T = T_0 + \eta \tag{6.39}$$

where u_i denotes the displacement components, T_0 denotes the reference temperature, and η denotes the small change in temperature from the reference temperature. For small deformations and temperature changes, we accordingly require that

$$\partial u_i / \partial X_j \ll 1 \tag{6.40}$$

$$\eta / T_0 \ll 1 \tag{6.41}$$

Under the condition of equation (6.40), the deformation tensor C_{ij} is given from Chapter 2 as

$$C_{ij} = \delta_{ij} + 2e_{ij} \tag{6.42}$$

where

$$e_{ij} = \frac{1}{2} \left(\frac{\partial u_i}{\partial X_j} + \frac{\partial u_j}{\partial X_i} \right) \tag{6.43}$$

denotes the usual strain tensor for small deformations. The rotation tensor R_{ij} is likewise expressible for small deformations as

$$R_{ij} = \delta_{ij} + w_{ij} \tag{6.44}$$

where

$$w_{ij} = \frac{1}{2} \left(\frac{\partial u_i}{\partial X_j} - \frac{\partial u_j}{\partial X_i} \right) \tag{6.45}$$

denotes the small-deformation rotation tensor.

Replacing C_{ij} with e_{ij} in the above constitutive relation for the Helmholtz function A and replacing T by η, we thus have, on expanding the function in a Taylor series about $e_{ij} = 0$ and $T = T_0$, the following expression accurate to second order in e_{ij} and η:

$$A(\eta, e_{ij}) = a + a_{ij} e_{ij} + b\eta + \tfrac{1}{2} a_{ijkl} e_{ij} e_{kl} + b_{ij} e_{ij} \eta + \tfrac{1}{2} \bar{c} \eta^2 \tag{6.46}$$

where a, a_{ij}, etc. denote constants.

The corresponding expressions for the entropy S and stress tensor t_{ij} follow immediately from equation (6.37). We have, in particular,

$$S = -b_{ij} e_{ij} - \bar{c} \eta - b \tag{6.47}$$

and

$$t_{ij} = \varrho_0 b_{ij} \eta + \varrho_0 a_{ijkl} e_{kl} \tag{6.48}$$

where, in view of equation (6.46), a_{ijkl} satisfies

$$a_{ijkl} = a_{klij} \tag{6.49}$$

and where a_{ij} has been set equal to zero in equation (6.46) in accordance with the requirement that the stress vanish when the strain does. Also, the current density ϱ has been replaced by the initial density ϱ_0 in equation (6.48) in accordance with the small-deformation assumption.

In addition to equations (6.47) and (6.48), we may also use equations (6.44) and (6.45) in the heat-flux expression of equation (6.37) to get, on expanding in a Taylor series about $e_{ij} = 0$, $T = T_0$,

$$q_i = a_i\eta + h_{ij}\frac{\partial\eta}{\partial X_j} + b_{ijk}e_{jk} \tag{6.50}$$

where a_i, h_{ij}, and b_{ijk} denote constants.

6.7. Restriction to Isotropic Materials

Equations (6.47), (6.48), and (6.50) contain various constants whose values depend on the particular material and material symmetries under consideration. If Q_{ij} denotes the components of rotations describing these symmetries, we require, in fact, from the arguments of Chapter 4 that the above constants a_i, b_{ij}, h_{ij}, b_{ijk}, and a_{ijkl} must satisfy

$$
\begin{aligned}
a_i &= Q_{ij}a_j \\
b_{ij} &= Q_{im}Q_{jn}b_{mn} \\
h_{ij} &= Q_{im}Q_{jn}h_{mn} \\
b_{ijk} &= Q_{im}Q_{jn}Q_{kp}b_{mnp} \\
a_{ijkl} &= Q_{im}Q_{jn}Q_{kp}Q_{lq}a_{mnpq}
\end{aligned}
\tag{6.51}
$$

Restricting attention to isotropic materials where all rotations are permissible symmetry rotations, it is easily seen from the arguments outlined in Chapter 4 that the above constants must reduce to the following form:

$$
\begin{aligned}
a_i &= 0 \\
b_{ij} &= \bar{a}\delta_{ij} \\
h_{ij} &= \bar{k}\delta_{ij} \\
b_{ijk} &= \bar{b}e_{ijk} \\
a_{ijkl} &= \bar{\lambda}\delta_{ij}\delta_{kl} + 2\bar{\mu}\delta_{ik}\delta_{jl}
\end{aligned}
\tag{6.52}
$$

where $\bar{\alpha}$, $\bar{k}$, $\bar{b}$, $\bar{\mu}$, and $\bar{\lambda}$ denote constants and δ_{ij} and e_{ijk} denote the Kronecker delta and permutation symbols as defined in Chapter 1.

Instead of using the above constants $\bar{\alpha}$, $\bar{k}$, etc., it is customary to introduce new constants defined by

$$\alpha = -\varrho_0\bar{\alpha}, \qquad k = -\bar{k}, \qquad \lambda = \varrho_0\bar{\lambda}, \qquad \mu = \varrho_0\bar{\mu}, \qquad c = -T_0\bar{c} \qquad (6.53)$$

where ϱ_0 denotes the initial density.

Substituting into equations (6.46)–(6.48) and (6.50), we find, with the help of the above equations, the following constitutive relations governing linearized thermoelastic deformations of an isotropic solid:

$$A = a + b\eta + \frac{1}{2}\,\frac{\lambda}{\varrho_0}\,(e_{kk})^2 + \frac{\mu}{\varrho_0}\,e_{kl}e_{kl} - \frac{\alpha}{\varrho_0}\,e_{kk}\eta - \frac{1}{2}\,\frac{c}{T_0}\,\eta^2 \quad (6.54)$$

$$S = \frac{\alpha}{\varrho_0}\,e_{kk} + \frac{c}{T_0}\,\eta - b \tag{6.55}$$

$$t_{ij} = (\lambda e_{kk} - \alpha\eta)\delta_{ij} + 2\mu e_{ij} \tag{6.56}$$

$$q_i = -k\,\frac{\partial\eta}{\partial X_i} \tag{6.57}$$

where, in obtaining equation (6.57) from (6.50) and (6.52), use has been made of the fact that e_{ij} is symmetric and, hence, that $e_{ijk}e_{jk} = 0$.

It should be noted that the second law as given by equation (6.21) is automatically satisfied by equation (6.57) for all $k \geq 0$ since these two equations combine to give for small deformations

$$q_i\,\frac{\partial T}{\partial x_i} = -k\,\frac{\partial\eta}{\partial X_i}\,\frac{\partial\eta}{\partial X_i} \tag{6.58}$$

which is always less than or equal to zero for all nonnegative k.

In the above equations, the constants λ and μ are the *Lamé constants* as first introduced in Chapter 4. The constant k is known as the *thermal conductivity* and the constant c is known as the *specific heat* of the material at constant volume. The ratio $\bar{\alpha} = \alpha/(3\lambda + 2\mu)$ is referred to as the *coefficient of thermal expansion*.

Table 6.1 gives typical values of these thermal constants for some common construction materials. Values of the elastic constants were given in Chapter 4.

TABLE 6.1. *Typical Values of Thermal Constants at Room Temperature*[a]

Material	ϱ_0, lbm/ft³	k, BTU/hr-ft-°F	c, BTU/lbm-°F	$\tilde{\alpha}$, 1/°F
Steels	4.87×10^2	2.5×10^1	0.11	0.7×10^{-5}
Aluminum	1.69×10^2	1.3×10^2	0.21	1.2×10^{-5}
Copper	5.59×10^2	2.2×10^2	0.09	0.9×10^{-5}
Glass	1.69×10^2	4.4×10^{-1}	0.20	0.4×10^{-5}

[a] Note: 1 lbm = (1/32.2) lb-sec²/ft = 0.454 kg.
    1 BTU = 778 ft-lb = 1.05 × 10³ N-m.

6.8. Governing Equations for Linear Thermoelastic Deformation of an Isotropic Solid

From Chapter 3, we may write the balance of linear and angular momentum corresponding to small deformations and displacements as

$$\varrho_0 \frac{\partial^2 u_i}{\partial t^2} = \frac{\partial t_{ji}}{\partial X_j} + f_i \tag{6.59}$$

$$t_{ji} = t_{ij} \tag{6.60}$$

To this same degree of approximation, the first law of thermodynamics, as given by equation (6.5), can also be written as

$$\varrho_0 \frac{\partial U}{\partial t} = \varrho_0 r - \frac{\partial q_i}{\partial X_i} + t_{ki} \frac{\partial v_i}{\partial X_k} \tag{6.61}$$

On writing the last term of this equation as

$$t_{ki} \frac{\partial v_i}{\partial X_k} = t_{ki} \frac{\partial^2 u_i}{\partial X_k \, \partial t} = t_{ki} \frac{\partial^2 u_i}{\partial t \, \partial X_k}$$

and expressing the displacement gradients in terms of the strain and rotation tensors through the relation

$$\frac{\partial u_i}{\partial X_k} = e_{ik} + w_{ik}$$

we find by direct expansion that

$$t_{ki} \frac{\partial v_i}{\partial X_k} = t_{ki} \frac{\partial e_{ik}}{\partial t}$$

Using this result and introducing the Helmholtz function A of equation (6.8), we thus find the first law expressible in the form

$$\varrho_0 \frac{\partial A}{\partial t} + \varrho_0 T_0 \frac{\partial S}{\partial t} + \varrho_0 S_0 \frac{\partial \eta}{\partial t} = \varrho_0 r - \frac{\partial q_i}{\partial X_i} + t_{ij} \frac{\partial e_{ij}}{\partial t} \quad (6.62)$$

where S_0 equals the value of the entropy when $\eta = 0$ and $e_{ij} = 0$.

On substituting the constitutive relations for an isotropic solid as given by equations (6.54)–(6.57) of the previous section, we have finally the following coupled equations governing linearized thermoelastic deformations of an isotropic solid:

$$\varrho_0 \frac{\partial^2 u_i}{\partial t^2} = f_i + (\lambda + \mu) \frac{\partial^2 u_k}{\partial X_i \, \partial X_k} + \mu \frac{\partial^2 u_i}{\partial X_k \, \partial X_k} - \alpha \frac{\partial \eta}{\partial X_i} \quad (6.63)$$

and

$$\varrho_0 c \frac{\partial \eta}{\partial t} + \alpha T_0 \frac{\partial^2 u_k}{\partial t \, \partial X_k} = \varrho_0 r + k \frac{\partial^2 \eta}{\partial X_k \, \partial X_k} \quad (6.64)$$

Selected Reading

Malvern, L. E., *Introduction to the Mechanics of a Continuous Medium*. Prentice-Hall, Englewood Cliffs, New Jersey, 1969. Chapter 5 discusses the first and second laws of thermodynamics as applied to a continuum body.

Boley, B. A., and J. H. Weiner, *Theory of Thermal Stresses*. John Wiley and Sons, New York, 1960. The equations of linear thermoelasticity are developed and applied to a number of important problems.

Coleman, B. D., and W. Noll, « The Thermodynamics of Elastic Materials with Heat Conduction and Viscosity », *Archive for Rational Mechanics and Analysis* **13**, 168–178 (1963). This paper presents the modern continuum treatment of thermal elasticity upon which the discussion of the present chapter is based.

Exercises

1. Using arguments similar to those required in Exercise 2 of Chapter 3, show from the energy balance relation of equation (6.4) that the following relation must apply at every point on a stationary shock surface:

    $$[\varrho(\tfrac{1}{2} v_i v_i + U) v_j - t_{ji} v_i + q_j] n_j = 0$$

    where brackets are used to denote the difference in the enclosed quantity immediately ahead of and behind the shock surface, and n_j denotes components of the normal to the shock surface.

2. Using the above result together with those obtained in Exercise 2 of Chapter 3, show that, for the one-dimensional case of a stationary plane compressive shock having unit normal in the x_1 direction, the following conditions must apply across the shock surface:

$$\varrho_A V_A = \varrho_B V_B$$

$$\varrho_A V_A(V_B - V_A) = P_A - P_B$$

$$\varrho_A V_A \left(\frac{V_B^2 - V_A^2}{2} + U_B - U_A \right) = P_A V_A - P_B V_B$$

where $t_{11} = -P$, $v_1 = V$, and the subscripts A and B denote conditions immediately ahead of and behind the shock surface.

Note that these relations can be applied to the case of a shock wave propagating with constant velocity V_s into stationary material by simply replacing V_A with $-V_s$ and V_B with $V_B - V_s$. The equations are then referred to as the *Rankine–Hugoniot relations* for a traveling shock front.

3. Show for small, isothermal (constant-temperature) deformations that the stress relation of equation (6.37) reduces to the strain-energy relation of equation (4.35) of Chapter 4.

4. Using equations (6.56) and (6.64), show for adiabatic (zero heat transfer) deformation that the isotropic constitutive relation of Chapter 4 applies provided the Lamé isothermal elastic constants λ and μ are replaced by corresponding *adiabatic elastic constants* λ' and μ', where

$$\lambda' = \lambda + \frac{\alpha^2 T_0}{\varrho_0 c}, \qquad \mu' = \mu$$

5. Using the constants given in Tables 4.1 and 6.1, show that the difference between adiabatic and isothermal elastic constants is very slight for ordinary materials like steel and aluminum.

7

Problems in Thermal Elasticity

In Chapter 6, we developed equations governing the elastic deformation of solids in the presence of thermal effects. We now apply this theory to a number of illustrative problems. As in our earlier work on elasticity in Chapter 5, we restrict attention primarily to the case of static deformations. At the outset, however, we do consider briefly the influence of thermal effects on the dynamic response of solids by examining a problem involving free vibrations.

7.1. Thermoelastic Vibrations

As a first example of the use of the thermoelastic equations of Chapter 6, we consider the idealized problem of free longitudinal vibration of a thermoelastic material of infinite extent. The problem is illustrated in Figure 7.1, where the horizontal coordinate X_1 has been taken as x and where, for initial conditions, we have assumed the particles of material to be displaced according to the relations

$$u_1(x, 0) = A \cos mx$$

$$\frac{\partial u_1(x, 0)}{\partial t} = 0 \tag{7.1}$$

where A and m denote constants.

To calculate the subsequent motion arising from the initial conditions, we consider the governing relations given by equations (6.63) and (6.64) of

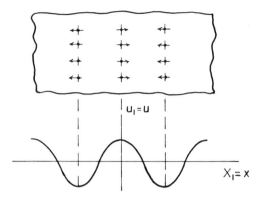

FIGURE 7.1.

the previous chapter. Assuming, in particular, a one-dimensional solution for the displacement u_i and the temperature increase η in the form

$$u_1 = u(x, t), \qquad u_2 = u_3 = 0$$
$$\eta = \eta(x, t) \tag{7.2}$$

these equations may be written for zero body forces and heat sources as

$$\frac{\partial^2 u}{\partial t^2} = C_1^2 \frac{\partial^2 u}{\partial x^2} - \gamma \frac{\partial \eta}{\partial x} \tag{7.3}$$

$$\frac{\partial \eta}{\partial t} + \beta \frac{\partial^2 u}{\partial x \, \partial t} = h \frac{\partial^2 \eta}{\partial x^2} \tag{7.4}$$

where C_1, γ, β, and h are defined in terms of the elastic constants λ and μ, the density ϱ_0, and the thermal constants α, c, and T_0 by the relations

$$C_1^2 = \frac{\lambda + 2\mu}{\varrho_0}, \qquad \gamma = \frac{\alpha}{\varrho_0}, \qquad \beta = \frac{\alpha T_0}{\varrho_0 c}, \qquad h = \frac{k}{\varrho_0 c} \tag{7.5}$$

In solving these equations subject to the initial conditions, we assume a solution for the displacement in the form

$$u = e^{-at}(A \cos \sigma t + B \sin \sigma t) \cos mx \tag{7.6}$$

where a and σ denote constants and where, in view of the initial conditions, B satisfies

$$B = \frac{a}{\sigma} A \tag{7.7}$$

Substituting this assumed solution into equation (7.3) and integrating, we find the associated increase in the temperature η above the reference temperature expressible in the form

$$\eta = e^{-at}(C \cos \sigma t + D \sin \sigma t) \sin mx \tag{7.8}$$

where the coefficients C and D are given by

$$C = -\frac{C_1^2 m^2 - a^2 - \sigma^2}{\gamma m} A$$

$$\tag{7.9}$$

$$D = -\frac{C_1^2 m^2 + a^2 + \sigma^2}{\gamma m} \frac{a}{\sigma} A$$

It remains to consider the energy relation as expressed by equation (7.4). Substituting equations (7.6) and (7.8) into this relation, we find the result expressible as

$$D_1 \cos \sigma t + D_2 \sin \sigma t = 0 \tag{7.10}$$

where D_1 and D_2 are given by

$$D_1 = -[2a^3 - 2a\sigma^2 + hm^2(C_1^2 m^2 - a^2 - \sigma^2)]$$

$$D_2 = \frac{a^2 + \sigma^2}{\gamma \sigma m} [a^2 - \sigma^2 + m^2(C_1^2 + \gamma\beta - ha)] - \frac{haC_1^2 m^3}{\gamma \sigma}$$

Now, the only way equation (7.10) can be satisfied for all times is for both D_1 and D_2 to vanish. Introducing the notation

$$\bar{\sigma} = \frac{\sigma}{mC_1}, \qquad \bar{a} = \frac{a}{mC_1}, \qquad \tau = \frac{\gamma\beta}{C_1^2}, \qquad \bar{h} = \frac{hm}{C_1} \tag{7.11}$$

we thus have the following relations for determining $\bar{a}$ and $\bar{\sigma}$:

$$\bar{\sigma}^2(\bar{h} - 2\bar{a}) - 2\bar{a}^3 + h\bar{a}^2 - \bar{h} = 0$$

$$(\bar{a}^2 + \bar{\sigma}^2)(\bar{a}^2 - \bar{\sigma}^2 + 1 + \tau - \bar{h}\bar{a}) - h\bar{a} = 0 \tag{7.12}$$

When the dimensionless ratio τ appearing in the second of these equations is set equal to zero, the solutions for $\bar{\sigma}$ and $\bar{a}$ are seen to be simply the isothermal elastic solutions, $\bar{\sigma} = 1$, $\bar{a} = 0$. For solids, this ratio of thermal constants is generally small, of the order of 10^{-2} or less, so that first-order solutions can therefore easily be determined as

$$\bar{a} = \frac{\bar{h}\tau}{2(1 + \bar{h}^2)}$$

$$\tag{7.13}$$

$$\bar{\sigma} = 1 + \frac{\tau}{2(1 + \bar{h}^2)}$$

or, in terms of the original variables, as

$$a = \frac{m^2 h \gamma \beta}{2(C_1^2 + h^2 m^2)}$$

$$\sigma = mC_1 \left[1 + \frac{\gamma \beta}{2(C_1^2 + h^2 m^2)} \right]$$

(7.14)

From the first of these solutions it can be seen that nonzero values of the thermal constants γ and β imply through equations (7.6) and (7.8) that the vibrations and nonuniform temperature distribution will be gradually damped out in the course of time. From the second of the solutions, it can also be seen that the presence of thermal effects is such as to cause the frequency of the vibrations to be increased over that which would exist if such effects were absent; that is, if $\gamma = \beta = 0$.

The physical explanation for the above *thermoelastic damping* of the vibrations is straightforward and centers mainly around the fact that any compression or expansion of a solid is associated with a corresponding rise or fall of its temperature. When these temperature changes are undisturbed, no thermal inteference with the motion will obviously arise. However, when heat conduction is present, as assumed in the above analysis, the temperature changes — and, hence, the motions — of the solid will be reduced during each compression or expansion.

7.2. Periodic Temperature Variation on the Boundary of a Thermoelastic Half-Space

We next consider the problem of a thermoelastic half-space subjected to a periodic temperature variation on its boundary. In particular, denoting the horizontal coordinate X_1 by x, we require the boundary at $x = 0$ to be free of applied forces and to be subjected to a temperature distribution of the form

$$\eta = \eta_0 \sin \sigma t$$

(7.15)

where η_0 and σ denote constants. The problem is illustrated in Figure 7.2.

The governing relations for this problem are given by equations (6.63) and (6.64) of Chapter 6. Assuming a one-dimensional solution for the displacement u_i and the temperature increase η of the form

$$u_1 = u(x, t), \qquad u_2 = u_3 = 0$$

$$\eta = \eta(x, t)$$

(7.16)

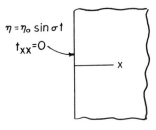

FIGURE 7.2.

we have, in particular,

$$\frac{\partial^2 u}{\partial t^2} = C_1^2 \frac{\partial^2 u}{\partial x^2} - \gamma \frac{\partial \eta}{\partial x} \tag{7.17}$$

$$\frac{\partial \eta}{\partial t} + \beta \frac{\partial^2 u}{\partial x \, \partial t} = h \frac{\partial^2 \eta}{\partial x^2} \tag{7.18}$$

where, as earlier, the constants C_1, γ, β, and h are defined in terms of the elastic constants, the density, and the thermal constants as

$$C_1^2 = \frac{\lambda + 2\mu}{\varrho_0}, \qquad \gamma = \frac{\alpha}{\varrho_0}, \qquad \beta = \frac{\alpha T_0}{\varrho_0 c}, \qquad h = \frac{k}{\varrho_0 c}$$

In order to simplify the problem as much as possible, we now make the plausible assumption (verified later) that the acceleration of the material resulting from the temperature variations will be insignificant, so that the term $\partial^2 u / \partial t^2$ may be ignored in the first of the above governing equations. We also assume tentatively that the change in temperature of the material arising from the accompaning change in strain will likewise be small, so that the term $\beta \, \partial^2 u / \partial x \, \partial t$ in the second of the above governing equations can also be ignored. Under these assumptions, equations (7.17) and (7.18) thus become expressible as

$$C_1^2 \frac{\partial^2 u}{\partial x^2} = \gamma \frac{\partial \eta}{\partial x} \tag{7.19}$$

$$\frac{\partial \eta}{\partial t} = h \frac{\partial^2 \eta}{\partial x^2} \tag{7.20}$$

The above assumptions leading to equations (7.19) and (7.20) are customarily employed in nonsteady heat flow problems of this kind, whether one, two, or three dimensional in nature, and give rise to what is referred to as *uncoupled quasistatic thermoelasticity theory.*

With equations (7.19) and (7.20) before us, we may now attempt to determine the temperature and displacement in the interior of the half-space under consideration. From the nature of the boundary condition, we may, in fact, expect the conditions in the interior to result from temperature waves emanating from the boundary and running forward into the solid at some finite speed. We accordingly assume a wave solution, consistent with the boundary condition, of the form

$$\eta = \eta_0 e^{-ax} \sin(\sigma t - mx) \tag{7.21}$$

where a and m denote constants.

Substituting this assumed solution into equation (7.20), we easily find the result expressible as

$$(\sigma - 2mah) \cos(\sigma t - mx) + h(m^2 - a^2) \sin(\sigma t - mx) = 0 \tag{7.22}$$

which requires, in turn, simply that m and a satisfy

$$m = a = \left(\frac{\sigma}{2h}\right)^{1/2} \tag{7.23}$$

With $m = a$ in equation (7.21), we may next calculate the displacement of particles resulting from this temperature distribution. From equation (7.19), we have, in particular, on integrating twice, that

$$u = -\frac{\gamma}{C_1^2} \frac{\eta_0}{2a} e^{-ax}[\sin(\sigma t - ax) - \cos(\sigma t - ax)] \tag{7.24}$$

where, in performing the integrations, we have used the conditions $u = 0$ when $x \to \infty$ and $t_{xx} = 0$ when $x = 0$, where t_{xx} is determined from equation (6.56) of Chapter 6.

Equations (7.21), (7.23), and (7.24) represent the solution to our problem as determined from the uncoupled quasistatic theory. It remains, however, to examine the consistency of this solution with the assumptions leading to such theory.

First consider the neglect of particle acceleration in equation (7.17). In order for the quasistatic assumption to be valid in the present problem, the acceleration determined from the above solution must be small in comparison with either of the two equal remaining terms in equation (7.17). Thus, we must have

$$\frac{\partial^2 u}{\partial t^2} \ll C_1^2 \frac{\partial^2 u}{\partial x^2} \tag{7.25}$$

Substituting the displacement solution of equation (7.24) into (7.25), we easily find that the quasistatic assumption requires therefore that

$$\sigma^2/2C_1^2 a \ll a$$

Or, on substituting for a from equation (7.23), that

$$\sigma \ll C_1^2/h \tag{7.26}$$

For most solids, the ratio on the right-hand side of this expression will be extremely large in comparison with any likely frequencies of interest, so that the quasistatic assumption can generally be expected to apply. For aluminum, for example, at room temperature, the ratio C_1^2/h is found to be approximately 3.8×10^{11} per second, so that frequencies of temperature variations as high as 10^9 per second would still allow the quasistatic assumption as a good approximation.

Next consider the uncoupling assumption involved in obtaining equation (7.20) from (7.18). For this assumption to be valid, we must require that

$$\beta \frac{\partial^2 u}{\partial x \, \partial t} \ll \frac{\partial \eta}{\partial t} \tag{7.27}$$

Using equations (7.24) and (7.21), we easily find that this condition requires that

$$\gamma \beta/C_1^2 \ll 1 \tag{7.28}$$

Recalling the definitions of γ, β, and C_1 given in connection with equations (7.17) and (7.18), this last condition can be written more explicitly as

$$\frac{\alpha^2 T_0}{\varrho_0 c} \frac{1}{\lambda + 2\mu} \ll 1 \tag{7.29}$$

For most solids at or near room temperature, the ratio of the constants on the left-hand side of this expression is small, of the order of 10^{-2} or less, so that the uncoupling assumption, like the quasistatic assumption, can generally be expected to apply. For the case of aluminum at room temperature, for example, the above ratio equals approximately 0.03.

7.3. Plane Strain and Plane Stress Thermoelastic Problems

The concept of plane strain, as introduced in Chapter 5, is often useful in examining the static or quasistatic deformation of long, cylindrical bodies

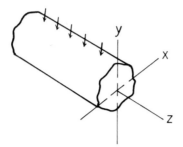

FIGURE 7.3.

having loads and thermal conditions uniform along the cylindrical axis, as shown in Figure 7.3.

In particular, we recall for the case of plane strain that the strain components e_{xz}, e_{yz}, and e_{zz} referred to axes $X_1 = x$, $X_2 = y$, $X_3 = z$ satisfy

$$e_{xz} = e_{yz} = e_{zz} = 0 \tag{7.30}$$

with the remaining strain components functions, at most, of the x and y in-plane coordinates. Under these conditions, the thermoelastic constitutive relations for an isotropic solid, as given by equation (6.56) of Chapter 6, require that

$$t_{yz} = t_{xz} = 0 \tag{7.31}$$

and

$$t_{zz} = \nu(t_{xx} + t_{yy}) - E\tilde{a}\eta \tag{7.32}$$

where E and ν denote Young's modulus and Poisson's ratio as defined in Chapter 4, and $\tilde{a} = \alpha/(3\lambda + 2\mu)$ denotes the coefficient of thermal expansion. With the help of equation (7.32), we can write the remaining in-plane constitutive relations as

$$e_{xx} = \frac{1+\nu}{E} [(1-\nu)t_{xx} - t_{yy}] + (1+\nu)\tilde{a}\eta \tag{7.33}$$

$$e_{yy} = \frac{1+\nu}{E} [(1-\nu)t_{yy} - \nu t_{xx}] + (1+\nu)\tilde{a}\eta \tag{7.34}$$

$$e_{xy} = \frac{1+\nu}{E} t_{xy} \tag{7.35}$$

Similarly, in dealing with thin plates or beams having in-plane edge loadings only and having thermal conditions constant through the thickness, as indicated in Figure 7.4, the concept of plane stress discussed in Chapter 5

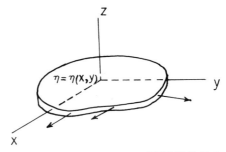

FIGURE 7.4.

is often useful. Plane stress, it will be recalled, is defined such that the stresses t_{xz}, t_{yz}, and t_{zz}, referred to axes $X_1 = x$, $X_2 = y$, $X_3 = z$, satisfy the relation

$$t_{xz} = t_{yz} = t_{zz} = 0 \qquad (7.36)$$

The constitutive relations for this case are given from equation (6.56) of Chapter 6 as

$$e_{xx} = \frac{1}{E}(t_{xx} - \nu t_{yy}) + \tilde{a}\eta$$

$$e_{yy} = \frac{1}{E}(t_{yy} - \nu t_{xx}) + \tilde{a}\eta$$

$$e_{zz} = -\frac{\nu}{E}(t_{xx} + t_{yy}) + \tilde{a}\eta \qquad (7.37)$$

$$e_{xy} = \frac{1+\nu}{E}t_{xy}$$

where, as before, E and ν denote Young's modulus and Poisson's ratio and $\tilde{a}$ denotes the coefficient of thermal expansion.

The stress equilibrium equations are identical for both plane strain and plane stress and take the form

$$\frac{\partial t_{xx}}{\partial x} + \frac{\partial t_{xy}}{\partial y} = 0 \qquad (7.38)$$

$$\frac{\partial t_{xy}}{\partial x} + \frac{\partial t_{yy}}{\partial y} = 0 \qquad (7.39)$$

where body forces have been assumed absent.

As in Chapter 5, these equations may automatically be satisfied by the introduction of an Airy stress function ϕ such that

$$t_{xx} = \frac{\partial^2 \phi}{\partial y^2}, \qquad t_{yy} = \frac{\partial^2 \phi}{\partial x^2}, \qquad t_{xy} = -\frac{\partial^2 \phi}{\partial x \, \partial y} \qquad (7.40$$

The compatibility relations of Chapter 5 need next to be examined. Again, as before, we have

$$\frac{\partial^2 e_{xx}}{\partial y^2} + \frac{\partial^2 e_{yy}}{\partial x^2} = 2\,\frac{\partial^2 e_{xy}}{\partial x\,\partial y} \tag{7.41}$$

Substituting the above constitutive relations into this equation and introducing the Airy stress function definitions, we thus find the governing equation for the stress function expressible as

$$\frac{\partial^4 \phi}{\partial x^4} + 2\,\frac{\partial^4 \phi}{\partial x^2\,\partial y^2} + \frac{\partial^4 \phi}{\partial y^4} = -\beta\left(\frac{\partial^2 \eta}{\partial x^2} + \frac{\partial^2 \eta}{\partial y^2}\right) \tag{7.42}$$

where β denotes a constant determined separately for plane strain and plane stress by

$$\beta = \frac{\tilde{a}E}{1-v} \qquad \text{plane strain}$$

$$\beta = \tilde{a}E, \qquad \text{plane stress}$$

From the energy relation of equation (6.64) of the previous chapter, we have also for the quasistatic uncoupled assumptions discussed in the last section that the temperature η must satisfy

$$\frac{\partial^2 \eta}{\partial x^2} + \frac{\partial^2 \eta}{\partial y^2} = \frac{\varrho_0 c}{k}\,\frac{\partial \eta}{\partial t} \tag{7.43}$$

where ϱ_0 denotes the density and c and k denote thermal constants. Substituting equation (7.43) into (7.42), we thus finally have the governing equation for the stress function expressible as

$$\frac{\partial^4 \phi}{\partial x^4} + 2\,\frac{\partial^4 \phi}{\partial x^2\,\partial y^2} + \frac{\partial^4 \phi}{\partial y^4} = -\beta\,\frac{\varrho_0 c}{k}\,\frac{\partial \eta}{\partial t} \tag{7.44}$$

Equations (7.43) and (7.44) provide governing thermoelastic equations for the temperature and stress function in two-dimensional static or quasi-static problems of plane strain and plane stress. These equations, together with suitable boundary conditions, may accordingly be used in place of the displacement and temperature relations of equations (6.63) and (6.64) of Chapter 6 for solving such problems. We recall, however, from Chapter 5 that, unlike plane strain, a two-dimensional treatment of plane stress is only approximate since compatibility relations in addition to equation (7.41) generally force a variation of the stresses across the thickness of the body.

For thin bodies, such variations will, of course, be small and the two-dimensional approximation may accordingly be regarded as good.

7.4. Thermal Stresses in a Thin Elastic Strip

As a simple example of the above concept of plane stress, we consider the problem of determining the thermal stresses in a heated strip of elastic material. The problem is illustrated in Figure 7.5. The strip is assumed infinitely long with stress-free longitudinal edges $y = \pm h$ insulated against any heat transfer.

For sake of simplicity, we assume the temperature distribution at time $t = 0$ to be given by

$$\eta = \eta_0 \cos mx \tag{7.45}$$

where η_0 and m denote constants. More complex initial conditions can, of course, be built up from this condition by Fourier synthesis.

The relation governing the temperature distribution at later times is given by equation (7.43) of the previous section. For temperature variations in the x direction only, such as involved here, this relation reduces to

$$\frac{\partial \eta}{\partial t} = \tilde{h} \frac{\partial^2 \eta}{\partial x^2} \tag{7.46}$$

where $\tilde{h} = k/\varrho_0 c$. The solution of this equation subject to the above initial conditions is easily seen to be expressible as

$$\eta = \eta_0 e^{-at} \cos mx \tag{7.47}$$

where $a = \tilde{h}m^2$.

From equation (7.44) of the previous section, we may next write the relation governing the stress function ϕ as

$$\frac{\partial^4 \phi}{\partial x^4} + 2 \frac{\partial^4 \phi}{\partial x^2 \partial y^2} + \frac{\partial^4 \phi}{\partial y^4} = \tilde{a} E m^2 \eta_0 e^{-at} \cos mx \tag{7.48}$$

To solve this equation, we assume a solution for ϕ of the form

$$\phi = e^{-at} f(y) \cos mx \tag{7.49}$$

which, on substitution, shows that the function $f(y)$ must satisfy

$$\frac{d^4 f}{dy^4} - 2m^2 \frac{d^2 f}{dy^2} + m^4 f = \tilde{a} E m^2 \eta_0 \tag{7.50}$$

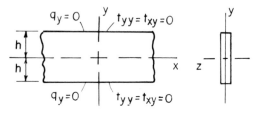

FIGURE 7.5.

This equation thus requires that $f(y)$ take the form

$$f = B \cosh my + Cy \sinh my + D \sinh my$$

$$+ Fy \cosh my + \frac{\tilde{a}E\eta_0}{m^2} \qquad (7.51)$$

where B, C, D, and F denote constants.

At the stress-free edges $y = \pm h$, the stresses t_{xy} and t_{yy} must both vanish. Remembering that

$$t_{xx} = \frac{\partial^2 \phi}{\partial y^2}, \qquad t_{yy} = \frac{\partial^2 \phi}{\partial x^2}, \qquad t_{xy} = -\frac{\partial^2 \phi}{\partial x\, \partial y} \qquad (7.52)$$

we find from equations (7.49) and (7.51) that the function f must satisfy the following conditions at these edges:

$$f = \frac{df}{dy} = 0 \qquad (7.53)$$

It is easily seen that these conditions require that the constants in equation (7.51) take the form

$$B = -\frac{\tilde{a}E\eta_0}{m^2} \frac{\sinh mh + mh \cosh mh}{\sinh mh \cosh mh + mh}$$

$$C = \frac{\tilde{a}E\eta_0}{m} \frac{\sinh mh}{\sinh mh \cosh mh + mh} \qquad (7.54)$$

$$D = F = 0$$

With the above constants, the stress function ϕ is accordingly established as

$$\phi = -\frac{\tilde{a}E\eta_0}{m^2} e^{-at} \left(\frac{\sinh mh + mh \cosh mh}{\sinh mh \cosh mh + mh} \cosh my \right.$$

$$\left. - \frac{m \sinh mh}{\sinh mh \cosh mh + mh} y \sinh my - 1 \right) \cos mx \qquad (7.55)$$

and the thermal stresses are thus finally determined from equation (7.52) as

$$t_{xx} = \tilde{a}E\eta_0 e^{-at} \left(\frac{\sinh mh - mh \cosh mh}{\sinh mh \cosh mh + mh} \cosh my \right.$$

$$+ \frac{m \sinh mh}{\sinh mh \cosh mh + mh} \; y \sinh my \right) \cos mx \qquad (7.56)$$

$$t_{yy} = \tilde{a}E\eta_0 e^{-at} \left(\frac{\sinh mh + mh \cosh mh}{\sinh mh \cosh mh + mh} \cosh my \right.$$

$$- \frac{m \sinh mh}{\sinh mh \cosh mh + mh} \; y \sinh my - 1 \right) \cos mx \qquad (7.57)$$

$$t_{xy} = \tilde{a}E\eta_0 e^{-at} \left(\frac{\sinh mh}{\sinh mh \cosh mh + mh} \; my \cosh my \right.$$

$$- \frac{mh \cosh mh}{\sinh mh \cosh mh + mh} \sinh my \right) \sin mx \qquad (7.58)$$

Application to Bars of Finite Length. The above solution may easily be applied to the determination of thermal stresses in bars of finite length (Figure 7.6). Assuming, for example, that the ends of the bar are free of applied forces and moments, we have, using Saint-Venant's principle, the following end boundary conditions applicable at $x = \pm L$:

$$F_x = b \int_{-h}^{+h} t_{xx} \, dy = 0 \qquad (7.59)$$

$$F_y = b \int_{-h}^{+h} t_{xy} \, dy = 0 \qquad (7.60)$$

$$M = b \int_{-h}^{+h} y t_{xx} \, dy = 0 \qquad (7.61)$$

Assuming, in addition, that the ends are maintained at the reference tem-

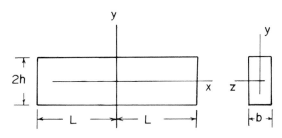

FIGURE 7.6.

perature, we also have at $x = \pm L$ that

$$\eta = 0 \tag{7.62}$$

Inspection of the above solution shows that all of these conditions can be satisfied by simply choosing m in the solution as $m = n\pi/2L$, where n denotes any odd integer. Moreover, for long bars where

$$mh = \frac{n\pi}{2}\,\frac{h}{L} \ll 1$$

the solution for the stresses may then be expressed approximately as

$$t_{xx} = \tilde{a}E\,\frac{m^2 y^2}{2}\,\eta_0 e^{-at}\cos mx$$

$$t_{yy} = t_{xy} \approx 0 \tag{7.63}$$

7.5. Plane Strain and Plane Stress Equations in Polar Coordinates

In terms of polar coordinates r, θ, equations (7.43) and (7.44) can be easily shown to be expressible as

$$\frac{\partial^2 \eta}{\partial r^2} + \frac{1}{r}\,\frac{\partial \eta}{\partial r} + \frac{1}{r^2}\,\frac{\partial^2 \eta}{\partial \theta^2} = \frac{\varrho_0 c}{k}\,\frac{\partial \eta}{\partial t} \tag{7.64}$$

and

$$\left(\frac{\partial^2}{\partial r^2} + \frac{1}{r}\,\frac{\partial}{\partial r} + \frac{1}{r^2}\,\frac{\partial^2}{\partial \theta^2}\right)\left(\frac{\partial^2 \phi}{\partial r^2} + \frac{1}{r}\,\frac{\partial \phi}{\partial r} + \frac{1}{r^2}\,\frac{\partial^2 \phi}{\partial \theta^2}\right)$$

$$= -\beta\,\frac{\varrho_0 c}{k}\,\frac{\partial \eta}{\partial t} \tag{7.65}$$

with the stresses being derived from the stress function according to the relations given in Chapter 5, namely

$$t_{rr} = \frac{1}{r}\,\frac{\partial \phi}{\partial r} + \frac{1}{r^2}\,\frac{\partial^2 \phi}{\partial \theta^2}$$

$$t_{\theta\theta} = \frac{\partial^2 \phi}{\partial r^2} \tag{7.66}$$

$$t_{r\theta} = \frac{1}{r^2}\,\frac{\partial \phi}{\partial \theta} - \frac{1}{r}\,\frac{\partial^2 \phi}{\partial r\,\partial \theta}$$

7.6. Hollow Circular Cylinder with Elevated Bore Temperature

As an example of the above concept of plane strain thermoelasticity, we consider the problem of determining the stresses in a long, hollow circular cylinder having a constant elevated temperature η_0 maintained on its interior surface. The problem is illustrated in Figure 7.7, where η has been taken equal to zero on the outside surface and where both surfaces have been assumed free of applied forces.

Assuming steady heat flow and axisymmetric solutions for the temperature η and the stress function ϕ, the governing relations, as given in polar coordinates by equations (7.64) and (7.65), become

$$\frac{d^2\eta}{dr^2} + \frac{1}{r}\frac{d\eta}{dr} = 0 \tag{7.67}$$

and

$$\frac{d^4\phi}{dr^4} + \frac{2}{r}\frac{d^3\phi}{dr^3} - \frac{1}{r^2}\frac{d^2\phi}{dr^2} + \frac{1}{r^3}\frac{d\phi}{dr} = 0 \tag{7.68}$$

The general solution of equation (7.67) is easily seen to be expressible as

$$\eta = G \log r + H \tag{7.69}$$

where G and H denote constants. Using the boundary conditions

$$\begin{aligned} \eta &= \eta_0 \quad \text{at} \quad r = a \\ \eta &= 0 \quad \text{at} \quad r = b \end{aligned} \tag{7.70}$$

we find immediately that

$$G = -\frac{\eta_0}{\log(b/a)} \tag{7.71}$$

$$H = \frac{\eta_0}{\log(b/a)} \log b$$

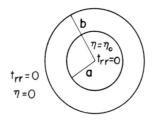

FIGURE 7.7.

so that the solution for η becomes

$$\eta = \frac{\eta_0}{\log(b/a)} \log \frac{b}{r} \tag{7.72}$$

The general solution of equation (7.68) for the stress function ϕ may also be seen to be expressible as

$$\phi = A \log r + Br^2 \log r + Cr^2 + D \tag{7.73}$$

where A, B, C, and D denote constants. The stresses are thus determined from equation (7.66) as

$$t_{rr} = \frac{A}{r^2} + B(1 + 2\log r) + 2C$$

$$t_{\theta\theta} = -\frac{A}{r^2} + B(3 + 2\log r) + 2C \tag{7.74}$$

$$t_{r\theta} = 0$$

The stress boundary conditions on the problem require for stress-free surfaces that

$$\hat{t}_r = t_{rr} = 0 \qquad \text{at} \quad r = a, b \tag{7.75}$$

Since there are three unspecified constants in equation (7.74) and only the two conditions of equation (7.75) to restrict them, the problem is thus seen to be indeterminate insofar as the stresses alone are concerned and we must thus consider also the displacements of the cylinder. This situation is, of course, analogous to that encountered in Chapter 5 when the stresses were examined in a hollow cylinder subjected to internal pressure.

In polar coordinates, the plane-strain consitutive relations analogous to equations (7.32)–(7.35) are expressible as

$$e_{rr} = \frac{1+v}{E} [(1-v)t_{rr} - vt_{\theta\theta}] + (1+v)\tilde{a}\eta$$

$$e_{\theta\theta} = \frac{1+v}{E} [(1-v)t_{\theta\theta} - vt_{rr}] + (1+v)\tilde{a}\eta \tag{7.76}$$

$$e_{r\theta} = \frac{1+v}{E} t_{r\theta}$$

$$t_{zz} = v(t_{rr} + t_{\theta\theta}) - E\tilde{a}\eta$$

The strain-displacement relations for axisymmetric displacements u_r, u_θ are given from Chapter 2 as

$$e_{rr} = \frac{du_r}{dr}$$

$$e_{\theta\theta} = \frac{u_r}{r} \tag{7.77}$$

$$2e_{r\theta} = \frac{du_\theta}{dr} - \frac{u_\theta}{r}$$

so that, on using these in equations (7.76), we have

$$\frac{du_r}{dr} = \frac{1+\nu}{E} [(1-\nu)t_{rr} - \nu t_{\theta\theta}] + (1+\nu)\tilde{\alpha}\eta \tag{7.78}$$

$$\frac{u_r}{r} = \frac{1+\nu}{E} [(1-\nu)t_{\theta\theta} - \nu t_{rr}] + (1+\nu)\tilde{\alpha}\eta \tag{7.79}$$

$$\frac{du_\theta}{dr} - \frac{u_\theta}{r} = 0 \tag{7.80}$$

Inserting the temperature and stress distributions given by equations (7.72) and (7.74) into the first of these relations, we find on integrating and comparing with the second that, for consistency, the constant B must be given by

$$B = \frac{E\tilde{\alpha}\eta_0}{4(1-\nu)\log(b/a)} \tag{7.81}$$

If the cylinder is assumed to be constrained against rigid-body rotations about its axes, equation (7.80) shows also that the displacement u_θ must vanish.

Returning to equation (7.74) and the two stress boundary conditions of equation (7.75), we may then find, for the above value of B, that the constants A and C must be given by

$$A = \frac{E\tilde{\alpha}\eta_0}{2(1-\nu)} \frac{a^2b^2}{b^2 - a^2}$$

$$2C = -\frac{E\tilde{\alpha}\eta_0}{4(1-\nu)\log(b/a)} \left[\frac{2a^2}{b^2 - a^2} \log\frac{b}{a} + 1 + 2\log b \right]$$

With the above constants, the stresses and radial displacement at all points in the cylinder are completely determined by equations (7.74) and

(7.79). In particular, the solution for the stresses t_{rr} and $t_{\theta\theta}$ may be written as

$$t_{rr} = \frac{\tilde{a}E\eta_0}{2(1-v)\log(b/a)}\left[-\log\frac{b}{r} - \frac{a^2}{b^2-a^2}\left(1-\frac{b^2}{r^2}\right)\log\frac{b}{a}\right]$$

(7.82)

$$t_{\theta\theta} = \frac{\tilde{a}E\eta_0}{2(1-v)\log(b/a)}\left[1-\log\frac{b}{r} - \frac{a^2}{b^2-a^2}\left(1+\frac{b^2}{r^2}\right)\log\frac{b}{a}\right]$$

(7.83)

The axial stress t_{zz} may also be determined from equation (7.76) as

$$t_{zz} = \frac{\tilde{a}E\eta_0}{2(1-v)\log(b/a)}\left[1 - 2\log\frac{b}{r} - \frac{2a^2}{b^2-a^2}\log\frac{b}{a}\right] \qquad (7.84)$$

7.7. Thermal Effects in Beam Deformations

In discussing the bending and extension of beams in the presence of thermal effects, we may use the framework of the approximate strength-of-material formulations given in Chapter 5. Consider, in particular, the initially straight beam of rectangular cross section shown in Figure 7.8. As indicated, we choose coordinates $X_1 = x$, $X_2 = y$ and take the origin of these coordinates such that the x axis lies along the initial centerline position of the beam.

Restricting attention to static deformations, we have the displacements $u_1 = u$, $u_2 = v$ at point x, y in the beam expressible from equations (5.143) of Chapter 5 as

$$u = u_0 - y\frac{dv_0}{dx}, \qquad v = v_0 \qquad (7.85)$$

where u_0 and v_0 denote horizontal and vertical displacements of the centerline

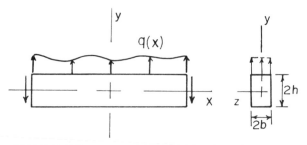

FIGURE 7.8.

of the beam. In addition, the equilibrium equations for the beam are also given from equations (5.146), (5.147), and (5.150) of Chapter 5 as

$$\frac{dN_{xx}}{dx} = 0 \tag{7.86}$$

$$\frac{dN_{xy}}{dx} + q(x) = 0 \tag{7.87}$$

$$\frac{dM_{xx}}{dx} - N_{xy} = 0 \tag{7.88}$$

where $q(x)$ denotes the lateral loading indicated in Figure 7.8 and where the stress resultants N_{xx} and N_{xy} and the moment M_{xx} are given in terms of the stresses and sectional area A by

$$N_{xx} = \int_A t_{xx}\, dA, \qquad N_{xy} = \int_A t_{xy}\, dA, \qquad M_{xx} = \int_A y t_{xx}\, dA \tag{7.89}$$

Modifying the constitutive relation of equation (5.142) of Chapter 5 to include thermal effects, we also have

$$t_{xx} = E e_{xx} - \tilde{a} E \eta \tag{7.90}$$

where E denotes Young's modulus, $\tilde{a}$ denotes the coefficient of thermal expansion, and η denotes the temperature increase. Using $e_{xx} = \partial u / \partial x$ and equation (7.85), we thus find from the above definitions of N_{xx} and M_{xx} that

$$N_{xx} = EA \frac{du_0}{dx} - N_T \tag{7.91}$$

$$M_{xx} = -EI \frac{d^2 v_0}{dx^2} - M_T \tag{7.92}$$

where I is determined as

$$I = \int_A y^2\, dA = \frac{4}{3} bh^3 \tag{7.93}$$

and N_T and M_T are given by

$$N_T = \int_A E\tilde{a}\eta\, dA$$

$$M_T = \int_A Ey\tilde{a}\eta\, dA \tag{7.94}$$

Finally, using equation (7.86) and eliminating N_{xy} between equations (7.87) and (7.88), we have with the help of the above expressions for N_{xx} and M_{xx} the following governing equations:

$$EA \frac{d^2u_0}{dx^2} = \frac{dN_T}{dx} \tag{7.95}$$

$$EI \frac{d^4v_0}{dx^4} = q - \frac{d^2M_T}{dx^2} \tag{7.96}$$

The boundary conditions to be used in connection with the above equations are similar to those discussed in Chapter 5. For the case of a *simply supported end*, we require at that end that

$$M_{xx} = -EI \frac{d^2v_0}{dx^2} - M_T = 0$$
$$\tag{7.97}$$
$$v_0 = 0$$

and either

$$u_0 = 0 \tag{7.98a}$$

or

$$\hat{N} = EA \frac{du_0}{dx} - N_T \tag{7.98b}$$

where $\hat{N}$ denotes an applied force, which may, of course, be set equal to zero for the case where end loadings are absent. For the case of a *built-in end*, we likewise require at that end that

$$v_0 = \frac{dv_0}{dx} = 0, \qquad u_0 = 0 \tag{7.99}$$

Application to Thermal Bending. We illustrate the above theory by considering the problem of a simply supported beam subjected to heating in such a way that the top surface is maintained at temperature $\eta = \eta_0$ and the bottom surface at temperature $\eta = 0$. The problem is illustrated in Figure 7.9.

Assuming the sides of the beam are insulated against heat transfer, we may suppose the temperature to be a function of the y coordinate only. The steady-state energy equation is thus determined from equation (6.64) of the preceding chapter as

$$\frac{d^2\eta}{dy^2} = 0 \tag{7.100}$$

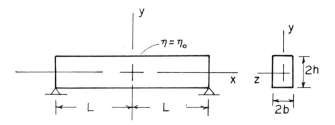

FIGURE 7.9.

and, in view of the boundary conditions, the temperature increase through-out the beam is therefore given by

$$\eta = \frac{\eta_0}{2h}(y + h) \tag{7.101}$$

From equation (7.96), we also have the governing relation for the vertical centerline deflection of the beam given for this case by

$$EI\frac{d^4v_0}{dx^4} = -\frac{d^2M_T}{dx^2} \tag{7.102}$$

where $I = 4bh^3/3$ and

$$M_T = \int_A Ey\bar{\alpha}\eta\,dA = \tfrac{2}{3}E\bar{\alpha}\eta_0bh^2 \tag{7.103}$$

Recognizing that M_T is independent of x, we thus have from equation (7.102) that

$$v_0 = c_0 + c_1x + c_2x^2 + c_3x^3 \tag{7.104}$$

where c_0, c_1, etc., denote constants. The boundary conditions of equation (7.97) require that at $x = \pm L$

$$v_0 = 0, \qquad \frac{d^2v_0}{dx^2} = -\frac{M_T}{EI} \tag{7.105}$$

so that the deflection is expressible as

$$v_0 = \frac{1}{2}\frac{M_T}{EI}(L^2 - x^2) = \frac{\bar{\alpha}\eta_0}{4h}(L^2 - x^2) \tag{7.106}$$

From equation (7.95), we also have that the horizontal deflection u_0 of the centerline of the beam must satisfy

$$EA\frac{d^2u_0}{dx^2} = \frac{dN_T}{dx} \tag{7.107}$$

where

$$N_T = \int_A E\tilde{a}\eta \, dA = 2E\tilde{a}\eta_0 \, hb \qquad (7.108)$$

Assuming boundary conditions such that $u_0 = 0$ at $x = \pm L$, we thus have from equation (7.107) that

$$u_0 = 0 \qquad (7.109)$$

everywhere along the beam.

With the above solutions for u_0 and v_0, we may easily determine the thermal stress t_{xx} in the beam from equations (7.85) and (7.90). We find

$$t_{xx} = \frac{M_T y}{I} - \tilde{a}E\eta = -\frac{E\tilde{a}\eta_0}{2} \qquad (7.110)$$

Using the equilibrium equations

$$\frac{\partial t_{xx}}{\partial x} + \frac{\partial t_{xy}}{\partial y} = 0$$

$$\frac{\partial t_{xy}}{\partial x} + \frac{\partial t_{yy}}{\partial y} = 0 \qquad (7.111)$$

with this result, we also find that

$$t_{xy} = t_{yy} = 0 \qquad (7.112)$$

everywhere in the beam.

Selected Reading

Boley, B. A., and J. H. Weiner, *Theory of Thermal Stresses*. John Wiley and Sons, New York, 1960. A number of technically important problems involving thermal elasticity are discussed in this book.

Timoshenko, S., and J. N. Goodier, *Theory of Elasticity*. McGraw-Hill Book Co., New York, 1951. Chapter 14 discusses thermal stress problems using quasistatic uncoupled equations of thermoelasticity.

Johns, D. J., *Thermal Stress Analysis*. Pergamon Press, New York, 1965. Applications of thermal elasticity to a number of practical problems are discussed.

Parkus, H., *Thermoelasticity*. Blaisdell Publishing Company, Waltham, Massachussetts, 1968. Advanced solution techniques are discussed.

Exercises

1. Using an approximate treatment paralleling that employed in Section 5.1, show that the governing thermoelastic equations for bars having stress-free insulated surfaces may be written as

$$\frac{\partial^2 u}{\partial t^2} = C_0^2 \frac{\partial^2 u}{\partial x^2} - \alpha^* \frac{\partial \eta}{\partial x}$$

$$\frac{\partial \eta}{\partial t} + \beta^* \frac{\partial^2 u}{\partial t \, \partial x} = k^* \frac{\partial^2 \eta}{\partial x^2}$$

where

$$C_0^2 = \frac{E}{\varrho_0}$$

$$\alpha^* = \frac{\mu}{\lambda + \mu} \frac{\alpha}{\varrho_0}$$

$$\beta^* = \frac{\mu \alpha T_0}{\varrho_0 c (\lambda + \mu) + \alpha^2 T_0}$$

$$k^* = \frac{(\lambda + \mu)k}{\varrho_0 c (\lambda + \mu) + \alpha^2 T_0}$$

2. Using the above results, determine the longitudinal displacement $u(x, t)$ in a thin rod of length L having a stress-free insulated surface and having stress-free ends maintained at the reference temperature $\eta = 0$. Assume initially that

$$u = A \cos \frac{n \pi x}{L}, \qquad \frac{\partial u}{\partial t} = 0$$

where A denotes a constant and n denotes an integer.

3. Consider a plate of thickness $2h$ and choose coordinate axes x and y in the midplane and z perpendicular. Assuming the plate surfaces and edges to be free of applied stress and that the temperature η and nonzero stresses t_{xx} and t_{yy} vary only through the thickness, show with the help of the compatibility equations of Section 5.4 and Saint-Venant's principle that

$$t_{xx} = t_{yy} = -\frac{\tilde{\alpha} E \eta}{1 - \nu} + \frac{\tilde{\alpha} E}{2h(1 - \nu)} \int_{-h}^{+h} \eta \, dz + \frac{3 \tilde{\alpha} E z}{2h^3 (1 - \nu)} \int_{-h}^{+h} z \eta \, dz$$

4. For plane-stress, axially symmetric deformations, show by combining

the appropriate equilibrium and constitutive relations that the radial displacement u must satisfy

$$\frac{d}{dr}\left[\frac{1}{r}\frac{d(ru)}{dr}\right] = (1+\nu)\bar{a}\frac{d\eta}{dr}$$

5. Show that the above result can be applied to plane-strain, axially symmetric deformations by simply replacing the coefficient $(1+\nu)$ by $(1+\nu)/(1-\nu)$.

6. Consider a thin circular disk of radius b having a concentric circular hole of radius a. Assuming the inside edge $(r = a)$ free of applied stress and maintained at temperature η_0 and the outside edge $(r = b)$ maintained at the reference temperature $\eta = 0$ and fixed against displacement, determine the displacement and stress distribution using the result of Exercise 4.

7. Reconsider the plate and shell theories of Sections 5.12 and 5.13 and incorporate thermal effects into the governing equations.

PART IV

VISCOUS ELASTICITY

The theory of viscous elasticity combines the simplest ideas of viscous fluids and elastic solids in an effort to account for certain effects, such as vibration damping and delayed elastic recovery, which are not predicted by elasticity theory alone.

The inclusion of simple one-dimensional viscous effects into elasticity theory is traceable to *Lord Kelvin* and his work on the decay of torsional oscillations in 1865. *Lord Rayleigh*, in his famous treatise, *Theory of Sound*, of 1877 also contributed to the theory of viscous elasticity by introducing general linear viscous terms into his treatments of the vibrations of elastic bodies. Further developments toward the end of the nineteenth century are due to *Woldemar Voigt*.

8

Theory of Viscous Elasticity

Earlier we examined the modifications of classical elasticity theory that arise when thermal effects are included in the formulation. We now proceed to consider briefly the corresponding modifications that result when *viscous* effects are included. These effects arise because the resistance to deformation of certain solids, called *viscoelastic solids*, is found to depend not only on the amount of elastic deformation, but also on the manner in which the deformation is produced. They are generally of minor importance in the deformation of common structural metals and alloys but have considerable significance when modern materials such as soft plastics or other polymers are considered.

8.1. Definition of a Standard Viscoelastic Solid

A *standard viscoelastic solid* is defined as one for which at each material point the following constitutive relation is taken to apply:

$$t_{ij} = \phi_{ij}\left(\frac{\partial x_k}{\partial X_p}, \frac{\partial v_k}{\partial x_p}, \dot{t}_{kp}\right) \tag{8.1}$$

where t_{ij} denotes components of the Cauchy stress tensor, $\dot{t}_{ij} = dt_{ij}/dt$ denotes the stress rate, ϕ_{ij} denotes an unspecified function, and x_k and v_k denote the position and velocity of a particle at time t which originally had coordinates X_p in the unstressed reference configuration of the solid.

8.2. Restrictions Placed by Principle of Material Indifference

The general nonlinear form of the viscoelastic solid defined by equation (8.1) must be restricted by the principle of material indifference to arbitrary, rigid, time-dependent motions as discussed in detail in Chapter 4. This principle requires, in particular, that if the constitutive relation of equation (8.1) is to hold for the motion defined by

$$x_k = x_k(X_p, t) \tag{8.2}$$

then the constitutive relation

$$t'_{ij} = \phi_{ij}\left(\frac{\partial x'_k}{\partial X_p}, \frac{\partial v'_k}{\partial x'_p}, i'_{kp} \right) \tag{8.3}$$

must likewise hold for the motion defined by

$$x'_k = Q_{km}(t)x_m + U_k(t) \tag{8.4}$$

where $Q_{km}(t)$ and $U_k(t)$ denote arbitrary, time-dependent, rigid-body rotations and translations of the solid and where

$$t'_{ij} = Q_{im}Q_{jn}t_{mn} \tag{8.5}$$

From this last equation, we have by differentiation that the stress rate in equation (8.3) is expressible as

$$i'_{kp} = Q_{km}Q_{pn}i_{mn} + (\dot{Q}_{km}Q_{pn} + Q_{km}\dot{Q}_{pn})t_{mn} \tag{8.6}$$

where, as in equation (8.1), a dot over a variable denotes a time derivative d/dt, holding initial coordinates fixed. Also, from equation (8.4), we have the deformation gradients and velocity gradients in equation (8.3) given by

$$\frac{\partial x'_k}{\partial X_p} = Q_{km} \frac{\partial x_m}{\partial X_p} \tag{8.7}$$

$$\frac{\partial v'_k}{\partial x'_p} = Q_{pr}\left(Q_{km} \frac{\partial v_m}{\partial x_r} + \dot{Q}_{kr} \right) \tag{8.8}$$

If we introduce components of the *rate of strain tensor* d_{ij} and the *rate of rotation tensor* Ω_{ij}, as defined by

$$d_{ij} = \frac{1}{2}\left(\frac{\partial v_i}{\partial x_j} + \frac{\partial v_j}{\partial x_i} \right) \tag{8.9}$$

$$\Omega_{ij} = \frac{1}{2}\left(\frac{\partial v_i}{\partial x_j} - \frac{\partial v_j}{\partial x_i} \right) \tag{8.10}$$

this last equation may be written alternatively as

$$\frac{\partial v_k'}{\partial x_p'} = Q_{pr}[Q_{km}(d_{mr} + \Omega_{mr}) + \dot{Q}_{kr}] \tag{8.11}$$

On substituting equations (8.5)–(8.7) and (8.11) into (8.3) and choosing

$$Q_{ij} = R_{ji}, \qquad \dot{Q}_{ij} = -Q_{im}\Omega_{mj} \tag{8.12}$$

where R_{ij} denotes the rotation associated with the motion defined by equation (8.2), we thus find the form of equation (8.1) satisfying the principle of material indifference expressible as

$$t_{rs} = R_{rm}R_{sn}\phi_{mn}(U_{kp}, d_{kp}^*, s_{kp}^*) \tag{8.13}$$

where U_{kp}, d_{kp}^*, and s_{kp}^* are defined as

$$U_{kp} = R_{mk}\frac{\partial x_m}{\partial X_p}$$

$$d_{kp}^* = R_{mk}R_{rp}d_{mr} \tag{8.14}$$

$$s_{kp}^* = R_{mk}R_{rp}s_{mr}$$

with the components of the so-called *corotational stress rate* s_{ij} defined by

$$s_{ij} = \dot{t}_{ij} - \Omega_{ir}t_{rj} - \Omega_{jr}t_{ir} \tag{8.15}$$

Equation (8.13) is analogous to that given by equation (4.15) of Chapter 4 for an elastic solid and becomes identical to it when the rate of strain d_{ij} and the stress rate s_{ij} are dropped from it.

8.3. Restriction to Small Deformations

In the case of small deformations, the rotation tensor R_{ij} and the stretch tensor U_{ij} are known from equations (2.47) and (2.48) of Chapter 2 to be expressible as

$$R_{ij} = \delta_{ij} + w_{ij} \tag{8.16}$$

$$U_{ij} = \delta_{ij} + e_{ij} \tag{8.17}$$

where w_{ij} and e_{ij} denote the small rotation and small strain tensors defined in terms of the displacement gradients by the usual relations

$$w_{ij} = \frac{1}{2}\left(\frac{\partial u_i}{\partial X_j} - \frac{\partial u_j}{\partial X_i}\right) \tag{8.18}$$

$$e_{ij} = \frac{1}{2}\left(\frac{\partial u_i}{\partial X_j} + \frac{\partial u_j}{\partial X_i}\right) \tag{8.19}$$

From equations (8.9), (8.14), and (8.15), we have also for small deformations that

$$d_{kp}^* = d_{kp} = \dot{e}_{kp} \tag{8.20}$$

$$s_{kp}^* = s_{kp} = \dot{t}_{kp} \tag{8.21}$$

Substituting the above equations into equation (8.13) and expanding, we find for small deformations that the constitutive relation must reduce to the following:

$$t_{rs} = \phi_{rs}(e_{kp}, \dot{e}_{kp}, \dot{t}_{kp}) \tag{8.22}$$

On expanding the above response function ϕ_{ij} into a Taylor series about $e_{kp} = \dot{e}_{kp} = \dot{t}_{kp} = 0$, we have further for small strains, small strain rates, and small stress rates that

$$t_{rs} = a_{rs} + B_{rspq}e_{pq} + C_{rspq}\dot{e}_{pq} - D_{rspq}\dot{t}_{pq} \tag{8.23}$$

where a_{rs}, B_{rspq}, etc. denote constants.

If the stress is assumed to vanish when the strain, strain rate, and stress rate vanish, the constant a_{rs} in the above equation must itself vanish. The constitutive relation thus becomes expressible finally as

$$t_{rs} = B_{rspq}e_{pq} + C_{rspq}\dot{e}_{pq} - D_{rspq}\dot{t}_{pq} \tag{8.24}$$

A solid obeying a constitutive relation of the above kind is often referred to as a *standard linear viscoelastic solid*. Also, if $D_{rspq} = 0$, the solid is referred to as a *Kelvin–Voigt viscoelastic solid*. Finally, if $B_{rspq} = 0$, we then have a so-called *Maxwell substance*.

8.4. Restriction to Isotropic Materials

The form of the material constants B_{rspq}, C_{rspq}, and D_{rspq} appearing in the above constitutive relation is restricted by the material symmetries existing in the particular solid under consideration. This matter has already been discussed in detail in Chapter 4 for the case of classical elasticity theory and the same general considerations may be employed here. In particular, if Q_{ij} denotes the components of rotations describing the particular material symmetries under consideration, we require from the arguments of Chapter 4 that these constants satisfy the relations

$$B_{rspq} = Q_{ri}Q_{sj}Q_{pk}Q_{ql}B_{ijkl}$$
$$C_{rspq} = Q_{ri}Q_{sj}Q_{pk}Q_{ql}C_{ijkl} \qquad (8.25)$$
$$D_{rspq} = Q_{ri}Q_{sj}Q_{pk}Q_{ql}D_{ijkl}$$

Restricting attention to isotropic materials where all rotations are permissible symmetry rotations, it is easily seen from the arguments of Chapter 4 that the constants B_{rspq}, C_{rspq}, and D_{rspq} must reduce, due to equation (8.25), to the following forms:

$$B_{rspq} = \lambda \, \delta_{rs} \delta_{pq} + 2\mu \, \delta_{rp} \delta_{sq}$$
$$C_{rspq} = \tilde{\lambda} \, \delta_{rs} \delta_{pq} + 2\tilde{\mu} \, \delta_{rp} \delta_{sq} \qquad (8.26)$$
$$D_{rspq} = \lambda^* \, \delta_{rs} \delta_{pq} + 2\mu^* \, \delta_{rp} \delta_{sq}$$

where λ and μ denote the usual Lamé elastic constants and where $\tilde{\lambda}$, $\tilde{\mu}$ and λ^*, μ^* denote corresponding viscous constants.

Substituting equation (8.26) into the constitutive relation of equation (8.24), we have, therefore, for the case of small deformation of an isotropic viscoelastic material, the following relation:

$$t_{rs} = \lambda e_{kk} \delta_{rs} + 2\mu e_{rs} + \tilde{\lambda} \dot{e}_{kk} \delta_{rs} + 2\tilde{\mu} \dot{e}_{rs} - \lambda^* \dot{t}_{kk} \delta_{rs} - 2\mu^* \dot{t}_{rs} \qquad (8.27)$$

8.5. Reduction of Constitutive Relations for Special Cases

Uniaxial Constitutive Relations. In discussing one-dimensional problems where only one nonzero stress component, say t_{xx}, exists (Figure 8.1), we have from equation (8.27) the following relations:

$$t_{xx} = (\lambda + 2\mu)e_{xx} + \lambda(e_{yy} + e_{zz})$$
$$+ (\tilde{\lambda} + 2\tilde{\mu})\dot{e}_{xx} + \tilde{\lambda}(\dot{e}_{yy} + \dot{e}_{zz}) - (\lambda^* + 2\mu^*)\dot{t}_{xx} \qquad (8.28)$$

$$t_{yy} = (\lambda + 2\mu)e_{yy} + \lambda(e_{xx} + e_{zz})$$
$$+ (\tilde{\lambda} + 2\tilde{\mu})\dot{e}_{yy} + \tilde{\lambda}(\dot{e}_{xx} + \dot{e}_{zz}) - \lambda^* \dot{t}_{xx} = 0 \qquad (8.29)$$

$$t_{zz} = (\lambda + 2\mu)e_{zz} + \lambda(e_{xx} + e_{yy})$$
$$+ (\tilde{\lambda} + 2\tilde{\mu})\dot{e}_{zz} + \tilde{\lambda}(\dot{e}_{xx} + \dot{e}_{yy}) - \lambda^* \dot{t}_{xx} = 0 \qquad (8.30)$$

t_{xx}

t_{xx}

FIGURE 8.1.

Noticing from the symmetry of the problem that

$$e_{yy} = e_{zz}, \qquad \dot{e}_{yy} = \dot{e}_{zz} \tag{8.31}$$

we may solve equation (8.29) or (8.30) for e_{yy} to get

$$e_{yy} = -\frac{\lambda}{2(\lambda+\mu)}\, e_{xx} - \frac{\tilde{\lambda}+\tilde{\mu}}{\lambda+\mu}\, \dot{e}_{yy} - \frac{\tilde{\lambda}}{2(\lambda+\mu)}\, \dot{e}_{xx} + \frac{\lambda^*}{2(\lambda+\mu)}\, i_{xx} \tag{8.32}$$

When this result is substituted into equation (8.28) and the symmetry relations of equation (8.31) are employed, we thus find

$$\begin{aligned}
t_{xx} = {} & \frac{\mu(3\lambda + 2\mu)}{\lambda + \mu}\, e_{xx} + \frac{\tilde{\lambda}\mu + 2\lambda\tilde{\mu} + 2\tilde{\mu}\mu}{\lambda + \mu}\, \dot{e}_{xx} \\
& - \frac{2(\lambda\tilde{\mu} - \tilde{\lambda}\mu)}{\lambda + \mu}\, \dot{e}_{yy} - \frac{\lambda^*\mu + 2\lambda\mu^* + 2\mu^*\mu}{\lambda + \mu}\, i_{xx}
\end{aligned} \tag{8.33}$$

The above relation shows that the uniaxial stress t_{xx} is dependent, in general, on the axial strain e_{xx} and strain rate $\dot{e}_{xx}$ as well as on the transverse strain rate $\dot{e}_{yy}$. If, however, the additional assumption is made that the viscoelastic constants $\tilde{\lambda}$ and $\tilde{\mu}$ are related to the elastic constants λ and μ by the relation

$$\frac{\tilde{\lambda}}{\tilde{\mu}} = \frac{\lambda}{\mu} \tag{8.34}$$

the term involving the transverse strain rate is then seen to vanish, and the resulting uniaxial constitutive relation thus becomes expressible simply as

$$t_{xx} = E e_{xx} + \tilde{E}\dot{e}_{xx} - E^* i_{xx} \tag{8.35}$$

where the constants E, $\tilde{E}$, and E^* are given by

$$E = \frac{\mu(3\lambda + 2\mu)}{\lambda + \mu}$$

$$\tilde{E} = \frac{\tilde{\lambda}\mu + 2\lambda\tilde{\mu} + 2\tilde{\mu}\mu}{\lambda + \mu}$$

$$E^* = \frac{\lambda^*\mu + 2\lambda\mu^* + 2\mu^*\mu}{\lambda + \mu}$$

Equation (8.35) is customarily employed in the one-dimensional stress analysis of viscoelastic solids. When viscous effects are absent, $\tilde{E} = E^* = 0$, and the equation is then seen to reduce simply to the usual one-dimensional elastic constitutive relation involving Young's modulus E. The limited generalization to include viscous effects in the form of equation (8.35) likens the one-dimensional response to that of a simple mechanical system having a series-connected elastic spring and viscous dashpot placed in parallel with a second elastic spring.

Inviscid Dilatational Response. A second case where the general constitutive relation of equation (8.27) can be simplified is that where the dilatational response of the solid is inviscid. From equation (8.27) we have, in particular, that

$$t_{kk} = (3\lambda + 2\mu)e_{kk} + (3\tilde{\lambda} + 2\tilde{\mu})\dot{e}_{kk} - (3\lambda^* + 2\mu^*)\dot{t}_{kk} \qquad (8.36)$$

where e_{kk} denotes the dilatation (or volume change) of an element of material. If this response is assumed inviscid, we must thus require that

$$\tilde{\lambda} = -\tfrac{2}{3}\tilde{\mu}, \qquad \lambda^* = -\tfrac{2}{3}\mu^* \qquad (8.37)$$

and the constitutive relations of equation (8.27) accordingly become expressible as

$$t_{rs} = \lambda e_{kk}\delta_{rs} + 2\mu e_{rs} + 2\tilde{\mu}(\dot{e}_{rs} - \tfrac{1}{3}\dot{e}_{kk}\delta_{rs}) - 2\mu^*(\dot{t}_{rs} - \tfrac{1}{3}\dot{t}_{kk}\delta_{rs}) \qquad (8.38)$$

8.6. Governing Equations for Linear Viscoelastic Deformation of an Isotropic Solid

We may combine the above general constitutive relation of equation (8.27) with the equations expressing the balance of linear momentum to obtain governing displacement equations for linear viscoelastic deformation of an isotropic solid. From Chapter 3, we have, in particular, the following equation expressing the balance of linear momentum:

$$\varrho_0 \frac{\partial^2 u_i}{\partial t^2} = \frac{\partial t_{ji}}{\partial X_j} + f_i \tag{8.39}$$

where ϱ_0 denotes the initial density of the material, u_i denotes the displacement components, and f_i denotes body-force components. Using these relations with equation (8.27) and the strain-displacement relation

$$e_{ij} = \frac{1}{2}\left(\frac{\partial u_i}{\partial X_j} + \frac{\partial u_j}{\partial X_i}\right) \tag{8.40}$$

we find immediately the governing equations:

$$\varrho_0 \ddot{u}_i = \lambda \frac{\partial^2 u_k}{\partial X_i \, \partial X_k} + \mu\left(\frac{\partial^2 u_j}{\partial X_j \, \partial X_i} + \frac{\partial^2 u_i}{\partial X_j \, \partial X_j}\right)$$

$$+ \tilde{\lambda}\frac{\partial^2 \dot{u}_k}{\partial X_i \, \partial X_k} + \tilde{\mu}\left(\frac{\partial^2 \dot{u}_j}{\partial X_j \, \partial X_i} + \frac{\partial^2 \dot{u}_i}{\partial X_j \, \partial X_j}\right)$$

$$- \lambda^* \frac{\partial i_{kk}}{\partial X_i} - 2\mu^* \frac{\partial i_{ri}}{\partial X_r} + f_i \tag{8.41}$$

$$\varrho_0 \ddot{u}_i = \frac{\partial i_{ji}}{\partial X_j} + \dot{f}_i \tag{8.42}$$

Selected Reading

Fung, Y. C., *Foundations of Solid Mechanics*. Prentice-Hall, Englewood Cliffs, New Jersey, 1965. Chapter 15 discusses viscoelasticity from a more general viewpoint than considered here.

Flugge, W., *Viscoelasticity*. Blaisdell Publishing Company, Waltham, Massachusetts, 1967. An introductory text dealing with both Kelvin–Voigt and Maxwell materials and their generalizations.

Lockett, F. J., *Nonlinear Viscoelastic Solids*. Academic Press, London, 1972. An advanced book treating general viscoelastic phenomena.

Exercises

1. Show that the choice of $\dot{Q}_{ij}$ in equation (8.12) is consistent with the orthogonality conditions

$$Q_{ik}Q_{jk} = Q_{ki}Q_{kj} = \delta_{ij}$$

2. Consider the rate of strain and rate of rotation components d_{ij} and Ω_{ij} as given by equations (8.9) and (8.10). If Δt denotes a small increment of time, show that $d_{ij}\Delta t$ and $\Omega_{ij}\Delta t$ can be regarded as measures of the strain and rotation of material at time $t + \Delta t$ relative to its configuration at time t.

3. The constitutive relations for one-dimensional elastic and viscous elements may be written as

$$\sigma = E\varepsilon, \qquad \sigma = C\dot{\varepsilon}$$

where $\sigma = t_{xx}$, $\varepsilon = e_{xx}$, and E and C denote constants. Show that a series connection of an elastic element of constant E_1 and a viscous element of constant C, when placed in parallel with an elastic element of constant E, gives rise to the response of equation (8.35) in the form

$$\sigma + \alpha\dot{\sigma} = E(\varepsilon + \beta\dot{\varepsilon})$$

where

$$\alpha = \frac{C}{E_1}, \qquad \beta = \frac{C(E + E_1)}{EE_1}$$

4. Determine the constitutive relation for uniaxial stress loading using the inviscid-dilatational-response relations of equations (8.36) and (8.38).

5. Reconsider Exercise 4 for the case of uniaxial strain.

6. Consider a small cylindrical element of viscoelastic material subjected to uniaxial loading governed by equation (8.35). Assuming the stress and strain to be of the form

$$t_{xx} = \sigma \sin \omega t, \qquad e_{xx} = \varepsilon \sin(\omega t - \alpha)$$

where σ, ε, ω, and α denote constants, determine the ratio σ/ε and α in terms of ω and the viscoelastic constants. The ratio σ/ε is referred to as the *dynamic modulus* and the constant α is referred to as the *viscous lag angle*.

9

Problems in Viscous Elasticity

We illustrate the theory of viscous elasticity in the present chapter by considering several basic problems involving either the standard linear viscoelastic solid, where the stress is dependent on the strain, strain rate, and the stress rate, or the Kelvin–Voigt viscoelastic solid, where the stress is dependent only on the strain and strain rate. Since, in either case, the viscoelastic equations reduce simply to the elastic relations of Chapter 4 when purely static deformations are considered, we shall limit our attention to dynamic or quasistatic problems, the latter being concerned, of course, with time-dependent deformations that occur slowly enough to allow the neglect of inertial effects.

9.1. Free Vibration of a Standard Viscoelastic Solid

We first consider the idealized problem of determining the motion of a viscoelastic material of infinite extent undergoing free transverse vibration. This problem is analogous to that considered in Chapter 7 in connection with longitudinal vibrations of a thermoelastic solid. As indicated in Figure 9.1, we choose x for the horizontal coordinate X_1 and assume for initial conditions that the particles of material are displaced according to the relations

$$u_2 = A \cos mx, \qquad \frac{\partial u_2}{\partial t} = 0 \tag{9.1}$$

where A and m denote constants.

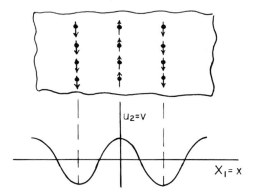

FIGURE 9.1.

To calculate the subsequent motion arising from these initial conditions, we assume a one-dimensional solution for the displacement in the form

$$u_2 = v(x, t), \qquad u_1 = u_3 = 0 \tag{9.2}$$

From the governing relations of equations (8.41) and (8.42) of Chapter 8, we then find for zero body forces that the function $v(x, t)$ must satisfy

$$2\mu^* \frac{\partial^3 v}{\partial t^3} + \frac{\partial^2 v}{\partial t^2} - D^2 \frac{\partial^3 v}{\partial t\, \partial x^2} - C^2 \frac{\partial^2 v}{\partial x^2} = 0 \tag{9.3}$$

where C^2 and D^2 are defined in terms of the elastic and viscous constants and the density as

$$C^2 = \mu/\varrho_0, \qquad D^2 = \tilde{\mu}/\varrho_0 \tag{9.4}$$

In order to solve equation (9.3) subject to the above initial conditions, we attempt a solution of the form

$$v = e^{-at}(A \cos \sigma t + B \sin \sigma t) \cos mx \tag{9.5}$$

where a and σ denote constants and where, due to the initial conditions, the constant B must be given by

$$B = \frac{a}{\sigma} A \tag{9.6}$$

in order to ensure no initial particle velocity.

Substituting equation (9.5) into equation (9.3), we find this assumed solution will, in fact, satisfy the governing displacement equation provided the constants a and σ are themselves solutions of the equations

$$[a^2 + \sigma^2][4a\mu^* - 1] + C^2 m^2 = 0 \qquad (9.7)$$

$$[a^2 + \sigma^2][2\mu^*(\sigma^2 - a^2) + a - D^2 m^2] + C^2 m^2 a = 0 \qquad (9.8)$$

When viscous effects are absent altogether ($\mu^* = D = 0$), these equations yield simply $a = 0$ and $\sigma = mC$. Hence, for small viscous effects, we may assume

$$a = \varepsilon_1, \qquad \sigma = mC(1 + \varepsilon_2) \qquad (9.9)$$

where ε_1 and ε_2 are small quantities, and easily determine first-order solutions in the form

$$
\begin{aligned}
a &= \frac{m^2(D^2 - 2\mu^* C^2)}{2 + 16(\mu^* mC)^2} \\
\sigma &= mC\left[\frac{1 + \mu^* m^2(6\mu^* C^2 + D^2)}{1 + 8(\mu^* mC)^2}\right]
\end{aligned}
\qquad (9.10)
$$

From the first of these solutions, we learn, using the definitions of C and D, that nonzero values of the viscous constants $\tilde{\mu}$ and μ^* imply through equation (9.5) that the vibrations will be damped out in the course of time, provided $\tilde{\mu} - 2\mu\mu^* > 0$. Moreover, when $16(\mu^* mC)^2 \gg 2$, the damping is seen to be essentially independent of the wave number m and, hence, of the frequency σ of the vibrations. From the second of the above solutions, we also learn that the presence of viscosity causes the frequency of the vibrations to be increased over that which would otherwise exist.

The above results are, of course, similar to those found in Chapter 7 for longitudinal vibrations of a thermoelastic solid. For the case of transverse vibrations such as considered here, no thermoelastic damping can, however, exist since the material experiences no compressions or expansions during the vibrations.

9.2. Time-Dependent Uniaxial Response of a Standard Viscoelastic Solid

We next consider the simple problem of determining the axial strain of a cylindrical specimen of viscoelastic material subjected to a time-dependent uniaxial stress loading (Figure 9.2).

Taking coordinate $X_1 = x$ along the axial direction of the cylinder and assuming the time dependence of the applied stress to be sufficiently weak

FIGURE 9.2.

to allow neglect of inertial effects, we may take the stress everywhere in the cylinder to be given by the simplified uniaxial constitutive relation of equation (8.35) of the preceding chapter, namely

$$t_{xx} = Ee_{xx} + \tilde{E}\dot{e}_{xx} - E^*\dot{t}_{xx} \tag{9.11}$$

With $t_{xx} = S(t)$, this equation may be written as

$$\frac{de_{xx}}{dt} + ae_{xx} = bS + c\dot{S} \tag{9.12}$$

where the constants a, b, and c are given by

$$a = E/\tilde{E}, \qquad b = 1/\tilde{E}, \qquad c = E^*/\tilde{E}$$

Integration accordingly yields

$$e_{xx} = \int_{-\infty}^{t} (bS + c\dot{S})e^{a(\tau-t)}\, d\tau \tag{9.13}$$

Equation (9.13) shows that the strain at time t depends on the entire history of the stress and stress rate. To examine the case where a constant stress S_0 is suddenly applied to a stress-free and strain-free specimen at time $t = 0$, we may assume the stress expressible for $t \geq 0$ as

$$S = S_0(1 - e^{-\alpha t}) \tag{9.14}$$

where α denotes a constant. Substituting this expression into equation (9.13) and integrating, we then find on letting α become infinitely large that

$$e_{xx} = \frac{S_0 b}{a}\left(1 - \frac{b - ca}{b}e^{-at}\right) \tag{9.15}$$

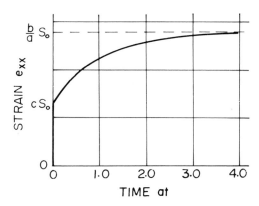

FIGURE 9.3.

which shows that, provided $a > 0$ and $b > ca$, the strain instantaneously jumps to the value cS_0 on applying the stress S_0 and that it subsequently increases to the final value S_0b/a. Such a response is referred to as a *creep response* (Figure 9.3).

Similarly, we may consider the case where a constant stress S_0 is applied for all times up to $t = 0$ and then suddenly reduced to zero. Immediately before releasing the stress, equation (9.13) shows, on carrying out the integration, that the strain is bS_0/a. After releasing it, we find on employing an integration procedure similar to that used above that the strain is given by

$$e_{xx} = \frac{S_0}{a} (b - ca)e^{-at} \qquad (9.16)$$

This result shows, for $a > 0$, $b > ca$, that immediately on releasing the

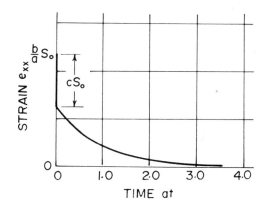

FIGURE 9.4.

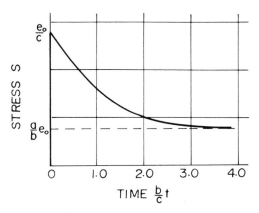

FIGURE 9.5.

stress, the strain drops by an amount cS_0 from its prior value of bS_0/a and then subsequently decays to zero. Such a response is called a *strain-relaxation response* (Figure 9.4).

In contrast with the case of applied stresses, we may consider alternatively the case where time-dependent strains are enforced on the specimen. The solution of equation (9.12) for the stress $S(t)$ then becomes

$$S = \int_{-\infty}^{t} \left(\frac{ae_{xx} - \dot{e}_{xx}}{c} \right) e^{(b/c)(\tau - t)} \, d\tau \tag{9.17}$$

Exactly as in the case of strain relaxation, we find the stress resulting from a suddenly enforced strain $e_{xx} = e_0$ at time $t = 0$ on a stress-free and strain-free specimen expressible as

$$S = \frac{ae_0}{b} + \frac{e_0}{c} \frac{b - ca}{b} e^{-(b/c)t} \tag{9.18}$$

This equation reveals, for $a > 0$, $b > ca$, that the stress jumps to the value e_0/c immediately after straining and that it then decays with time to the value ae_0/b. Such a response is referred to as a *stress-relaxation response* (Figure 9.5).

9.3. Hollow Circular Cylinder of Kelvin–Voigt Material Subjected to Periodic Bore Pressure

As a further illustration of the governing viscoelastic equations of Chapter 8, we next consider the problem of determining the displacement and stresses in a long, hollow, circular cylinder of Kelvin–Voigt material having a periodic pressure loading applied to its inside surface. The problem is illustrated in Figure 9.6. Here $t_{rr}(a) = -P_0 \sin \sigma t$ and $t_{rr}(b) = 0$.

Taking displacements in polar coordinates such that

$$u_r = u_r(r, t), \qquad u_\theta = u_z = 0 \tag{9.19}$$

we have from Chapter 2 that the only nonzero strains are the polar strains e_{rr} and $e_{\theta\theta}$, and these are given by

$$e_{rr} = \frac{\partial u_r}{\partial r}, \qquad e_{\theta\theta} = \frac{u_r}{r} \tag{9.20}$$

The constitutive relations connecting the normal stresses t_{rr} and $t_{\theta\theta}$ with the above strains for a Kelvin–Voigt material are given from equation (8.27) of the preceding chapter (with $\lambda^* = \mu^* = 0$) as

$$t_{rr} = (\lambda + 2\mu)e_{rr} + \lambda e_{\theta\theta} + (\tilde{\lambda} + 2\tilde{\mu})\dot{e}_{rr} + \tilde{\lambda}\dot{e}_{\theta\theta} \tag{9.21}$$

$$t_{\theta\theta} = (\lambda + 2\mu)e_{\theta\theta} + \lambda e_{rr} + (\tilde{\lambda} + 2\tilde{\mu})\dot{e}_{\theta\theta} + \tilde{\lambda}\dot{e}_{rr} \tag{9.22}$$

Also, assuming the frequency σ of the applied pressure loading to be sufficiently low to allow the neglect of inertial effects, we have the equation expressing the balance of linear momentum in the radial direction given from Chapter 3 as

$$\frac{\partial t_{rr}}{\partial r} + \frac{t_{rr} - t_{\theta\theta}}{r} = 0 \tag{9.23}$$

On combining equations (9.20)–(9.23), we thus find the equation governing radial displacement in the cylinder expressible as

$$\frac{\partial^2 u_r}{\partial r^2} + \frac{1}{r}\frac{\partial u_r}{\partial r} - \frac{u_r}{r} + \beta\left(\frac{\partial^3 u_r}{\partial r^2 \partial t} + \frac{1}{r}\frac{\partial^2 u_r}{\partial r \partial t} - \frac{1}{r^2}\frac{\partial u_r}{\partial t}\right) = 0 \tag{9.24}$$

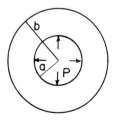

FIGURE 9.6.

where β is given by

$$\beta = \frac{\tilde{\lambda} + 2\tilde{\mu}}{\lambda + 2\mu} \tag{9.25}$$

In view of the assumed periodic pressure loading indicated in Figure 9.6, we assume a solution for the radial displacement of the form

$$u_r = f(r) \sin \sigma t + g(r) \cos \sigma t \tag{9.26}$$

where $f(r)$ and $g(r)$ denote functions to be determined and where σ denotes the specified frequency of the applied pressure loading.

On substituting equation (9.26) into the governing relation of equation (9.24), we find the assumed solution acceptable provided the functions $f(r)$ and $g(r)$ satisfy

$$\frac{d^2 f}{dr^2} + \frac{1}{r} \frac{df}{dr} - \frac{1}{r^2} f = 0 \tag{9.27}$$

$$\frac{d^2 g}{dr^2} + \frac{1}{r} \frac{dg}{dr} - \frac{1}{r^2} g = 0 \tag{9.28}$$

It is a simple matter to show that these equations accordingly require that $f(r)$ and $g(r)$ take the form

$$f = Ar + \frac{B}{r} \tag{9.29}$$

$$g = Cr + \frac{D}{r} \tag{9.30}$$

where A, B, C, and D denote constants. The displacement u_r of equation (9.26) may thus be written as

$$u_r = \left(Ar + \frac{B}{r} \right) \sin \sigma t + \left(Cr + \frac{D}{r} \right) \cos \sigma t \tag{9.31}$$

To determine the constants in this expression, we must use the boundary conditions on the stress, namely that

$$\hat{t}_r = t_{rr} = -P_0 \sin \sigma t \qquad \text{at} \quad r = a$$
$$\hat{t}_r = t_{rr} = 0 \qquad\qquad \text{at} \quad r = b \tag{9.32}$$

On combining equations (9.20), (9.21) and (9.31), and using these boundary conditions, it can be found after some reduction that the constants A, B, C, and D must be given by the relations

$$A = \frac{P_0}{2} \frac{\lambda + 2\mu}{(\lambda + \mu)^2 + \sigma^2(\tilde{\lambda} + \tilde{\mu})^2} \frac{a^2}{b^2 - a^2}$$

$$B = \frac{P_0}{2} \frac{\mu}{\mu^2 + \sigma^2\tilde{\mu}^2} \frac{a^2 b^2}{b^2 - a^2}$$

$$C = -\frac{P_0}{2} \frac{\sigma(\tilde{\lambda} + \tilde{\mu})}{(\lambda + \mu)^2 + \sigma^2(\tilde{\lambda} + \tilde{\mu})^2} \frac{a^2}{b^2 - a^2}$$

$$D = -\frac{P_0}{2} \frac{\sigma\tilde{\mu}}{\mu^2 + \sigma^2\tilde{\mu}^2} \frac{a^2 b^2}{b^2 - a^2}$$

These constants, together with equations (9.31), (9.20), and (9.21), thus provide the solution for the displacement and stresses in the cylinder. For the radial displacement, we find, in particular, that

$$u_r = \frac{P_0}{2} \frac{a^2 r}{b^2 - a^2} \frac{(\lambda + \mu) \sin \sigma t - \sigma(\tilde{\lambda} + \tilde{\mu}) \cos \sigma t}{(\lambda + \mu)^2 + \sigma^2(\tilde{\lambda} + \tilde{\mu})^2}$$
$$+ \frac{P_0}{2} \frac{a^2 b^2}{b^2 - a^2} \frac{1}{r} \frac{\mu \sin \sigma t - \sigma\tilde{\mu} \cos \sigma t}{\mu^2 + \sigma^2\tilde{\mu}^2} \tag{9.33}$$

and, for the radial and circumferential stresses, that

$$t_{rr} = \frac{a^2 P_0}{b^2 - a^2} \left(1 - \frac{b^2}{r^2} \right) \sin \sigma t$$
$$t_{\theta\theta} = \frac{a^2 P_0}{b^2 - a^2} \left(1 + \frac{b^2}{r^2} \right) \sin \sigma t \tag{9.34}$$

It is interesting to note that, since the above stress relations do not contain any elastic or viscous constants, they are independent of the degree of viscosity actually present and are therefore applicable also for the case of purely elastic deformations. This may be seen more directly by simply replacing $P_0 \sin \sigma t$ with P in the above solution and comparing it with the corresponding elastic solution given by equation (5.127) of Chapter 5.

9.4. Viscous Effects in Beam Deformations

In discussing the bending and extension of beams made of viscoelastic material, we may again use the framework of the strength-of-materials formulation given in Chapter 5. Consider, for example, an initially straight beam of Kelvin–Voigt material such as shown in Figure 9.7. As indicated,

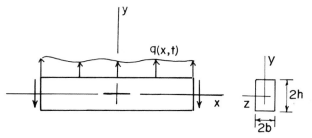

FIGURE 9.7.

we choose coordinates $X_1 = x$, $X_2 = y$ and take the origin such that the x axis lies along the initial centerline of the beam.

Restricting attention to quasistatic deformations where inertial effects may be neglected, we have the displacements $u_1 = u$, $u_2 = v$ at point x, y in the beam expressible from equation (5.143) of Chapter 5 as

$$u = u_0 - y\frac{\partial v_0}{\partial x}, \qquad v = v_0 \tag{9.35}$$

where u_0 and v_0 denote the horizontal and vertical displacements of the centerline of the beam. In addition, the equilibrium equations for the beam are given from equations (5.146), (5.147), and (5.150) of Chapter 5 as

$$\frac{\partial N_{xx}}{\partial x} = 0 \tag{9.36}$$

$$\frac{\partial N_{xy}}{\partial x} + q(x, t) = 0 \tag{9.37}$$

$$\frac{\partial M_{xx}}{\partial x} - N_{xy} = 0 \tag{9.38}$$

where $q(x, t)$ denotes the applied lateral loading as shown in Figure 9.7 and where N_{xx}, N_{xy}, and M_{xx} are given by

$$N_{xx} = \int_A t_{xx}\, dA, \qquad N_{xy} = \int_A t_{xy}\, dA, \qquad M_{xx} = \int_A y t_{xx}\, dA \tag{9.39}$$

with A denoting the cross-sectional area of the beam.

Modifying the constitutive relation of equation (5.142) of Chapter 5 to include viscous effects, we also have, for the restrictive uniaxial constitutive relation of equation (8.35) of the preceding chapter (with $E^* = 0$) that

$$t_{xx} = E e_{xx} + \tilde{E}\dot{e}_{xx} \tag{9.40}$$

where E denotes Young's modulus and $\tilde{E}$ denotes the viscous modulus.

Using $e_{xx} = \partial u/\partial x$ and equation (9.35), we may easily find from the above definitions of N_{xx} and M_{xx} that

$$N_{xx} = EA \frac{\partial u_0}{\partial x} + \tilde{E}A \frac{\partial^2 u_0}{\partial x \, \partial t} \tag{9.41}$$

$$M_{xx} = -EI \frac{\partial^2 v_0}{\partial x^2} - \tilde{E}I \frac{\partial^3 v_0}{\partial x^2 \, \partial t} \tag{9.42}$$

where I is given by

$$I = \int_A y^2 \, dA$$

Finally, using equations (9.36)–(9.38) with the above results, we find the governing equations for centerline displacement of the beam expressible as

$$EA \frac{\partial^2 u_0}{\partial x^2} + \tilde{E}A \frac{\partial^3 u_0}{\partial x^2 \, dt} = 0 \tag{9.43}$$

and

$$EI \frac{\partial^4 v_0}{\partial x^4} + \tilde{E}I \frac{\partial^5 v_0}{\partial x^4 \, \partial t} = q(x, t) \tag{9.44}$$

The boundary conditions to be used with the above equations are similar to those discussed in Chapter 5. For a *simply supported end*, we require at that end that

$$M_{xx} = -EI \frac{\partial^2 v_0}{\partial x^2} - \tilde{E}I \frac{\partial^3 v_0}{\partial x^2 \, \partial t} = 0$$
$$v_0 = 0 \tag{9.45}$$

and either

$$u_0 = 0 \tag{9.46}$$

or

$$\hat{N} = EA \frac{\partial u_0}{\partial x} + \tilde{E}A \frac{\partial^2 u_0}{\partial x \, \partial t} \tag{9.47}$$

where $\hat{N}$ denotes an applied axial loading. Similarly, for a *built-in end* we require at that end that

$$v_0 = \frac{\partial v_0}{\partial x} = 0, \qquad u_0 = 0 \tag{9.48}$$

Application to Uniform Loading. As an illustration of the above equations, we consider the simple problem of a simply supported beam of

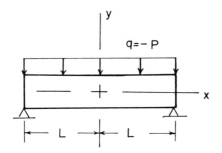

FIGURE 9.8.

Kelvin–Voigt viscoelastic material subjected at time $t = 0$ to a uniform loading $q = -p$ (Figure 9.8).

From equation (9.44) we have the governing equation for the centerline deflection v_0 expressible as

$$\frac{\partial}{\partial t}\left(\frac{\partial^4 v_0}{\partial x^4}\right) + \frac{E}{\tilde{E}} \frac{\partial^4 v_0}{\partial x^4} = -\frac{p}{\tilde{E}I} \tag{9.49}$$

Integrating with respect to time, we find

$$\frac{\partial^4 v_0}{\partial x^4} = -\frac{p}{EI}(1 - e^{-(E/\tilde{E})t}) \tag{9.50}$$

which may then be integrated with respect to x to obtain

$$v_0 = \left(-\frac{px^4}{24EI} + \frac{c_1 x^3}{6} + \frac{c_2 x^2}{2} + c_3 x + c_4\right)(1 - e^{-(E/\tilde{E})t}) \tag{9.51}$$

where c_1, c_2, etc. denote constants. Using the boundary conditions of equation (9.45) at $x = \pm L$, we easily find that

$$c_2 = \frac{pL^2}{2EI}, \qquad c_4 = -\frac{5}{24}\frac{pL^4}{EI}, \qquad c_1 = c_3 = 0 \tag{9.52}$$

so that the solution for v_0 is expressible as

$$v_0 = -\frac{pL^4}{24EI}\left(\frac{x^4}{L^4} - \frac{6x^2}{L^2} + 5\right)(1 - e^{-(E/\tilde{E})t}) \tag{9.53}$$

Assuming, in addition, that the longitudinal displacement $u_0 = 0$ at the ends $x = \pm L$, we also have from equation (9.43) that

$$u_0 = 0 \tag{9.54}$$

everywhere along the beam.

With the above solutions for v_0 and u_0, we may easily determine the longitudinal stress in the beam from equations (9.35) and (9.40). We find, in particular, that

$$t_{xx} = \frac{py}{I}(x^2 - L^2) \tag{9.55}$$

Using the stress-equilibrium equations as in Section 5.11 of Chapter 5, we also find the stresses t_{xy} and t_{yy} expressible as

$$t_{xy} = -\frac{px}{2I}(y^2 - h^2) \tag{9.56}$$

$$t_{yy} = \frac{p}{2I}\left(\frac{y^3}{3} - h^2 y - \frac{2h^3}{2}\right) \tag{9.57}$$

It is worth noting that, since the above stresses are independent of time, they have the same values regardless of whether the beam is in its initial, straight position (at $t = 0$) or in its fully deflected position (at $t = \infty$).

9.5. Viscoelastic Correspondence Principle

In considering complex quasistatic viscoelastic solutions, it is sometimes convenient to make use of what is called the *correspondence principle*. This principle arises because of the similarity which exists between Laplace transforms with respect to time of the governing viscoelastic equations and the corresponding untransformed elastic equations.

To see this similarity, we have only to examine the two sets of equations. Remembering that the Laplace transform $\bar{f}(X_i, s)$ with respect to time of a variable $f(X_i, t)$ is defined as

$$\bar{f}(X_i, s) = \int_0^\infty f(X_i, t)e^{-st}\, dt \tag{9.58}$$

we have, on multiplying the equilibrium equations

$$\frac{\partial t_{ji}}{\partial X_j} + f_i = 0 \tag{9.59}$$

through by e^{-st} and integrating each term over the indicated limits, the following *transformed* equation:

$$\frac{\partial \bar{t}_{ji}}{\partial X_j} + \bar{f}_i = 0 \tag{9.60}$$

Similarly, the strain-displacement relations

$$2e_{ij} = \frac{\partial u_i}{\partial X_j} + \frac{\partial u_j}{\partial X_i} \tag{9.61}$$

are transformed as

$$2\bar{e}_{ij} = \frac{\partial \bar{u}_i}{\partial X_j} + \frac{\partial \bar{u}_j}{\partial X_i} \tag{9.62}$$

In addition, if we introduce the *deviatoric stress* p_{ij} and the *deviatoric strain* g_{ij} defined by

$$\begin{aligned} p_{ij} &= t_{ij} - \tfrac{1}{3} t_{kk}\, \delta_{ij} \\ g_{ij} &= e_{ij} - \tfrac{1}{3} e_{kk}\, \delta_{ij} \end{aligned} \tag{9.63}$$

we may write the viscoelastic constitutive relations of equation (8.35) of the preceding chapter as

$$\begin{aligned} p_{ij} &= 2\mu g_{ij} + 2\tilde{\mu}\dot{g}_{ij} - 2\mu^* \dot{p}_{ij} \\ t_{kk} &= B e_{kk} + \tilde{B}\dot{e}_{kk} - B^* \dot{t}_{kk} \end{aligned} \tag{9.64}$$

where

$$B = 3\lambda + 2\mu, \qquad \tilde{B} = 3\tilde{\lambda} + 2\tilde{\mu}, \qquad B^* = 3\lambda^* + 2\mu^*$$

and these may be transformed to yield

$$\begin{aligned} \bar{p}_{ij} &= \frac{2(\mu + s\tilde{\mu})}{1 + 2s\mu^*}\, \bar{g}_{ij} \\ \bar{t}_{kk} &= \frac{B + s\tilde{B}}{1 + sB^*}\, \bar{e}_{kk} \end{aligned} \tag{9.65}$$

provided $t_{ij} = e_{ij} = 0$ at $t = 0$.

Equations (9.60), (9.62), and (9.65), together with similarly transformed boundary conditions of stress t_{ij} or displacement u_i, are easily seen from Chapters 4 and 5 to be identical to the corresponding untransformed equations governing elastic deformations provided, in the elastic constitutive relations,

$$p_{ij} = 2\mu g_{ik}, \qquad t_{kk} = B e_{kk} \tag{9.66}$$

the constants μ and B are replaced by μ' and B', where

$$\mu' = \frac{\mu + s\tilde{\mu}}{1 + 2s\mu^*}, \qquad B' = \frac{B + s\tilde{B}}{1 + sB^*} \tag{9.67}$$

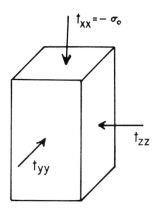

FIGURE 9.9.

If we have an elasticity solution for the stresses and displacements in a problem of interest, we may therefore use the above correspondence to write down immediately the Laplace transforms of the associated viscoelastic solution. In particular, if the elastic solution is expressed in terms of the elastic constants λ and μ, we have only to substitute $\lambda = (B - 2\mu)/3$ and then replace B and μ with μ' and B' as given by equation (9.67). By inverting these transformed equations, we may thus finally obtain equations describing the viscoelastic stresses and displacements.

9.6. Laterally Constrained Bar

As a simple illustration of the above correspondence principle, we consider the problem of determining the lateral stresses necessary to prevent lateral strains in a Kelvin–Voigt viscoelastic bar subjected at time $t = 0$ to a constant compressive axial stress $t_{xx} = -\sigma_0$. The problem is illustrated in Figure 9.9.

In keeping with the correspondence principle, we first solve for the elastic lateral stresses $t_{yy} = t_{zz}$ necessary to maintain lateral strains.

From the elastic constitutive relations, we have

$$t_{xx} = (\lambda + 2\mu)e_{xx}$$
$$t_{yy} = t_{zz} = \lambda e_{xx}$$

(9.68)

so that with $t_{xx} = -\sigma_0$, the stress t_{yy} is given by

$$t_{yy} = -\frac{\lambda \sigma_0}{\lambda + 2\mu} - \frac{(B - 2\mu)\sigma_0}{B + 4\mu}$$

(9.69)

Now, by the correspondence principle, the Laplace transform of the lateral stress t_{yy} in a viscoelastic bar, having stress $t_{xx} = -\sigma_0$ applied at time $t = 0$ is expressible as

$$\bar{t}_{yy} = \frac{(B + 2\mu) + s(\tilde{B} + 2\tilde{\mu})}{B + 4\mu + s(\tilde{B} + 4\tilde{\mu})} \,\bar{\sigma}_0 \tag{9.70}$$

or, since $\bar{\sigma}_0 = \sigma_0/s$, this may be written as

$$t_{yy} = -\frac{\lambda\sigma_0}{(a + bs)s} - \frac{\tilde{\lambda}\sigma_0}{a + bs} \tag{9.71}$$

where $a = \lambda + 2\mu$ and $b = \tilde{\lambda} + 2\tilde{\mu}$, and where the relations $B = 3\lambda + 2\mu$ and $\tilde{B} = 3\tilde{\lambda} + 2\tilde{\mu}$ have been used.

It is easy to see that the inverse transform of the first term in this expression is

$$\frac{\lambda\sigma_0}{a} \left(e^{-(a/b)t} - 1\right)$$

and that the inverse of the second term is

$$\frac{\tilde{\lambda}\sigma_0}{b} e^{-(a/b)t}$$

Hence, since the Laplace transform is linear, we thus have the lateral stress for the viscoelastic solid expressible as

$$t_{yy} = -\frac{\lambda\sigma_0}{a} + \frac{\lambda\sigma_0}{a}\left(1 + \frac{\tilde{\lambda}a}{b}\right)e^{-(a/b)t} \tag{9.72}$$

Note that for large times, the stress t_{yy} is approximately

$$t_{yy} = -\frac{\lambda\sigma_0}{a} = -\frac{\lambda\sigma_0}{\lambda + 2\mu} \tag{9.73}$$

which is identical to that provided by the elastic solution. The above solution thus represents a creep response similar to that discussed in Section 9.2.

Selected Reading

Flugge, W., *Viscoelasticity*. Blaisdell Publishing Company, Waltham, Massachusetts, 1967.
    Solutions to a number of problems involving viscoelastic response are considered.
Fung, Y. C., *Foundations of Solid Mechanics*. Prentice-Hall, Englewood Cliffs, New Jersey,
    1965. Chapters 1 and 15 give brief discussions of theory and problems in viscoelas-
    ticity.

Exercises

1. Using an approximate treatment paralleling that employed in Section 5.1, show that the governing viscoelastic equation for prismatic bars having stress-free lateral surfaces may be written as

$$\varrho_0 E^* \frac{\partial^3 u}{\partial t^3} + \varrho_0 \frac{\partial^2 u}{\partial t^2} - \tilde{E} \frac{\partial^3 u}{\partial t \, \partial x^2} - E \frac{\partial^2 u}{\partial x^2} = 0$$

where u denotes longitudinal displacement.

2. Using the above equation, determine the longitudinal displacement of a bar of length L having stress-free ends and subject to the initial conditions

$$u = A \cos \frac{n\pi x}{L}, \qquad \frac{\partial u}{\partial t} = 0$$

where n denotes any integer.

3. Show in the above exercise that the condition for the amplitude of the vibrations to remain small is

$$\tilde{E} - EE^* \geq 0$$

This result, as well as the corresponding one found in Section 9.1, is referred to as a *material stability condition* since, if the condition did not apply, a small initial displacement disturbance would increase in amplitude during the vibrations.

4. Consider a Kelvin–Voigt viscoelastic solid of infinite extent and analyze the propagation of longitudinal waves of the form

$$u = e^{-ax} \cos(kx - \omega t)$$

where a, k, and ω denote constants.

5. Reconsider Exercise 4 for the case of transverse waves.

PART V

PLASTICITY

The theory of plasticity is concerned with the mechanics of solids, particularly metals, that are stressed to such a level that deformations occur that remain present after the removal of all loads.

The theory had its beginning in 1864, when *H. Tresca* reported experiments to the French Academy which suggested that the plastic yielding of a metal occurs when the maximum shear stress reaches a critical value. Shortly thereafter, the mathematician *Barre de Saint-Venant* took up the problem of plastic flow and developed governing two-dimensional equations which were subsequently generalized to three dimensions by *M. Levy* in 1871. No further significant advances in the theory occurred until 1913, when *R. von Mises* published his researches on the subject in the German literature. In place of *Tresca's* yield condition, he proposed on mathematical grounds a yield criterion which was later shown to be superior to *Tresca's* in describing experimental yield measurements and to be equivalent to an elastic energy condition suggested by *James Maxwell* in 1856. He also proposed as plastic flow relations the equations developed earlier by *Levy*. This theory, together with generalizations to include elastic strains in the flow equations by *L. Prandtl* in 1924 and *E. Reuss* in 1930, constitutes essentially the theory of plasticity as used today.

10

Theory of Plasticity

In the previous discussions of elasticity theory, it was assumed that the mechanical response of the solid was such that all strains vanished with the vanishing of all externally applied disturbances. From simple experiments on certain solids, particularly structural metals, it has, however, been found that this purely elastic response breaks down at some critical stress level and that the recoverable small elastic deformations are then accompanied by *plastic deformations* which remain present after all external disturbances vanish. In an effort to understand this phenomenon of plastic distortion, we accordingly consider the theory of plasticity in the present chapter, regarding it—as in the case of thermal elasticity and viscous elasticity—primarily as a supplement to the theory of elasticity described in Chapter 4.

10.1. Definition of an Elastic-Plastic Solid

An *elastic-plastic solid* is defined as one for which, at each material point, the purely elastic response of the material breaks down when the Cauchy stress components satisfy a *yield condition* of the form

$$f(t_{ij}) = 0 \tag{10.1}$$

and for which the subsequent deformation is governed by constitutive relations of the form

$$\dot{t}_{ij} = \phi_{ij}\left(\frac{\partial v_p}{\partial x_k}, t_{pk}\right), \qquad f = 0$$

$$\dot{t}_{ij} = \phi_{ij}\left(\frac{\partial v_p}{\partial x_k}\right), \qquad f < 0 \tag{10.2}$$

where f and ϕ_{ij} denote unspecified functions, t_{ij} denote the Cauchy stress components, $\dot{t}_{ij} = dt_{ij}/dt$ denotes stress-rate components, and v_i denotes velocity components of the particle at place x_i and time t.

10.2. Restrictions Placed by Principle of Material Indifference

As in previous work, we require that if equations (10.1) and (10.2) are to hold for the motion described by

$$x_i = x_i(X_j, t) \tag{10.3}$$

then the relations

$$f(t'_{ij}) = 0 \tag{10.4}$$

and

$$\dot{t}'_{ij} = \phi_{ij}\left(\frac{\partial v'_p}{\partial x'_k}, t'_{pk}\right), \qquad f = 0$$

$$\dot{t}'_{ij} = \phi_{ij}\left(\frac{\partial v'_p}{\partial x'_k}\right), \qquad f < 0 \tag{10.5}$$

must also hold for the motion

$$x'_i = Q_{ij}x_j + U_i \tag{10.6}$$

where Q_{ij} and U_i denote an arbitrary rigid rotation and translation of the solid and where, as in Chapter 7,

$$t'_{ij} = Q_{im}Q_{jn}t_{mn} \tag{10.7}$$

$$\dot{t}'_{ij} = Q_{im}Q_{jn}\dot{t}_{mn} + (\dot{Q}_{im}Q_{jn} + Q_{im}\dot{Q}_{jn})t_{mn} \tag{10.8}$$

$$\frac{\partial v'_i}{\partial x'_j} = Q_{jn}\left(Q_{im}\frac{\partial v_m}{\partial x_n} + \dot{Q}_{in}\right) \tag{10.9}$$

Yield Condition. Considering first the yield condition, we have, on substituting equation (10.6) into (10.4), that the yield condition must be expressible as

$$f(Q_{im}Q_{jn}t_{mn}) = 0 \tag{10.10}$$

Choosing Q_{ij} so as to provide principal stresses t'_{11}, t'_{22}, t'_{33}, this equation becomes

$$f(t'_{11}, t'_{22}, t'_{33}) = 0 \tag{10.11}$$

Alternatively, since the principal stresses are determined by the *stress invariants*

$$I_1 = t_{ii}$$
$$I_2 = \tfrac{1}{2}t_{ii}t_{jj} - \tfrac{1}{2}t_{ij}t_{ij} \tag{10.12}$$
$$I_3 = e_{ijk}t_{i1}t_{j2}t_{k3}$$

as given by the coefficients of equation (1.39) of Chapter 1, equation (10.5) may also be written as

$$f(I_1, I_2, I_3) = 0 \tag{10.13}$$

which is easily seen to satisfy the principle of material indifference.

Constitutive Relations. Next consider the elastic-plastic constitutive relations. If, as in Chapter 7, we introduce components of the rate of strain tensor d_{ij} and the rate of rotation tensor Ω_{ij}, as defined by

$$d_{ij} = \frac{1}{2}\left(\frac{\partial v_i}{\partial x_j} + \frac{\partial v_j}{\partial x_i}\right)$$
$$\Omega_{ij} = \frac{1}{2}\left(\frac{\partial v_i}{\partial x_j} - \frac{\partial v_j}{\partial x_i}\right) \tag{10.14}$$

equation (10.9) may be written as

$$\frac{\partial v_i'}{\partial x_j'} = Q_{jn}[Q_{im}(d_{mn} + \Omega_{mn}) + \dot{Q}_{in}] \tag{10.15}$$

Hence, on substituting this equation into equation (10.5) and choosing

$$\dot{Q}_{ij} = -Q_{im}\Omega_{mj} \tag{10.16}$$

we find the reduced constitutive relations

$$s_{ij}' = \phi_{ij}(d_{pk}', t_{pk}'), \qquad f = 0$$
$$s_{ij}' = \phi_{ij}(d_{pk}'), \qquad f < 0 \tag{10.17}$$

where

$$s_{ij}' = Q_{im}Q_{jn}s_{mn}, \qquad d_{ij}' = Q_{im}Q_{jn}d_{mn} \tag{10.18}$$

and where s_{ij} denotes components of the corotational stress rate, defined as in Chapter 7, as

$$s_{ij} = \dot{t}_{ij} - \Omega_{ir}t_{rj} - \Omega_{jr}t_{ir} \tag{10.19}$$

Since equation (10.17) is of the same form for any rotation, we may obviously choose $Q_{ij} = \delta_{ij}$ to obtain reduced constitutive relations satisfying the principle of material indifference in the form

$$
\begin{aligned}
s_{ij} &= \phi_{ij}(d_{pk}, t_{pk}), & f &= 0 \\
s_{ij} &= \phi_{ij}(d_{pk}), & f &< 0
\end{aligned}
\tag{10.20}
$$

10.3. Restriction to Quasilinear Response Independent of Mean Stress

Yield Condition. Equation (10.13) may be regarded as the general form of the yield condition satisfying the principle of material indifference. When, as customarily assumed, the yielding is also supposed insensitive to changes in the mean stress $t_{kk}/3$, as achieved, for example, by the addition of a hydrostatic pressure loading, the yield condition is further restricted such that, if equation (10.13) is to hold for one stress state, it must also hold for any other stress state differing only by the mean stress. Choosing, in particular, the stress state $\tilde{t}_{ij}$ having zero mean stress, equation (10.13) thus becomes expressible as

$$
f(J_2, J_3) = 0
\tag{10.21}
$$

where

$$
J_2 = \tfrac{1}{2}\tilde{t}_{ij}\tilde{t}_{ij}, \qquad J_3 = e_{ijk}\tilde{t}_{i1}\tilde{t}_{j2}\tilde{t}_{k3}
$$

On expanding equation (10.21) in a Taylor series and retaining only the term of lowest order in the stress, we immediately obtain the *Maxwell–Mises yield condition* in the form

$$
J_2 = \tfrac{1}{2}\tilde{t}_{ij}\tilde{t}_{ij} = K^2
\tag{10.22}
$$

where K denotes a constant. Substituting the total stress $t_{ij} = \tilde{t}_{ij} + \bar{t}\,\delta_{ij}$, where $\bar{t} = t_{kk}/3$, this relation is also expressible as

$$
J_2 = \tfrac{1}{2}(t_{ij} - \bar{t}\,\delta_{ij})(t_{ij} - \bar{t}\,\delta_{ij}) = K^2
\tag{10.23}
$$

Finally, considering the case of uniaxial loading with only the stress component, say, t_{11} nonzero, we find

$$
K = Y/\sqrt{3}
\tag{10.24}
$$

where Y denotes the yield stress in uniaxial loading.

It is worth noting in connection with the above work that an alternate yield condition, based on the maximum shear stress and associated with the name of Tresca, is sometimes employed in place of or as an approximation to the Maxwell–Mises condition. Experiments, particularly on structural metals, generally favor, however, the Maxwell–Mises condition.

Constitutive Relations. If we restrict attention to the case where the constitutive relations of equation (10.20) are expressible in the quasilinear form

$$s_{ij} = C_{ijkl}d_{kl} + D_{ijkl}t_{kl} \tag{10.25}$$

where

$$D_{ijkl} = D_{ijkl}, \qquad f = 0$$
$$D_{ijkl} = 0, \qquad f < 0$$

and where C_{ijkl} and D_{ijkl} are functions, at most, of invariants formed from the stress and rate of strain, we have immediately from equation (10.17) that C_{ijkl} and D_{ijkl} must satisfy

$$C_{ijkl} = Q_{pi}Q_{qj}Q_{mk}Q_{nl} \quad C_{pqmn}$$
$$D_{ijkl} = Q_{pi}Q_{qj}Q_{mk}Q_{nl} \quad D_{pqmn} \tag{10.26}$$

and, hence, must be of the form of equation (4.52) of Chapter 4, namely

$$C_{ijkl} = \lambda \, \delta_{ij} \, \delta_{kl} + 2\mu \, \delta_{ik} \, \delta_{jl}$$
$$D_{ijkl} = \psi \, \delta_{ij} \, \delta_{kl} + 2\phi \, \delta_{ik} \, \delta_{jl} \tag{10.27}$$

where λ, μ, ψ, and ϕ denote scalar functions of invariants formed from the stress and rate of strain.

Restricting further to the case where λ and μ are constants and where the material is insensitive to changes in the mean stress $\bar{t} = t_{kk}/3$, we easily find, on adopting the Maxwell–Mises yield condition, a properly invariant form of the *Prandtl–Reuss elastic-plastic constitutive relations* expressible as

$$s_{ij} = \lambda d_{kk} \, \delta_{ij} + 2\mu d_{ij} + 2\phi(t_{ij} - \bar{t} \, \delta_{ij}) \tag{10.28}$$

where

$$\phi = \phi, \qquad J_2 = K^2$$
$$\phi = 0, \qquad J_2 < K^2$$

and where λ and μ are taken as the Lamé elastic constants for consistency with the time derivative of the linear elastic equations of Chapter 4.

An alternate form of the above constitutive relations is found by solving for the rate of strain in terms of the stress and stress rate. We find

$$d_{ij} = \frac{1 + v}{E} s_{ij} - \frac{v}{E} s_{kk} \delta_{ij} + \frac{\beta}{K} (t_{ij} - \bar{t} \delta_{ij}) \qquad (10.29)$$

where

$$\beta = \beta, \qquad J_2 = K^2$$
$$\beta = 0, \qquad J_2 < K^2$$

and where E and v denote Young's modulus and Poisson's ratio and β is related to ϕ through the equation

$$\beta = - \frac{2K(1 + v)}{E} \phi \qquad (10.30)$$

10.4. Plastic Constitutive Relations Applicable for Negligible Elastic Deformations

When plastic deformation of a solid can proceed freely in at least one direction, the rate of plastic strain will generally be much greater than the associated rate of elastic strain after only a small amount of actual plastic deformation. In these circumstances, the Prandtl–Reuss elastic-plastic constitutive relations can be simplified by neglecting altogether the elastic contribution to the total strain rate. As an approximation to the full con-stitutive relations of equation (10.29), we then have

$$d_{ij} = \frac{\beta}{K} (t_{ij} - \bar{t} \delta_{ij}) \qquad (10.31)$$

where

$$\beta = \beta, \qquad J_2 = K^2$$
$$\beta = 0, \qquad J_2 < K^2$$

These relations, which apply for rigid-plastic solids, are referred to as the *Levy–Mises constitutive relations.*

If we solve for $t_{ij} - \bar{t} \delta_{ij}$ using equation (10.31) and substitute the result into the Maxwell–Mises yield condition, we find for the case of negligible elastic deformation that β is expressible simply as

$$\beta = (\tfrac{1}{2} d_{ij} d_{ij})^{1/2} \qquad (10.32)$$

when $J_2 = K^2$.

10.5. Governing Equations

In considering practical problems in plasticity, we are usually not concerned entirely with small deformations, so that use must generally be made of the nonlinear forms of the equations expressing the balance of linear and angular momentum, the conservation of mass, and the constitutive material response. From equations (3.35) and (3.39) of Chapter 3, we have, in particular, the balance of linear and angular momentum given as

$$\varrho \, \frac{dv_i}{dt} = \frac{\partial t_{ji}}{\partial x_j} + f_i \tag{10.33}$$

$$t_{ij} = t_{ji} \tag{10.34}$$

where ϱ denotes the current density of the material, v_i denotes the velocity, t_{ij} denotes the Cauchy stress, and f_i denotes the body force.

From equation (3.31) of Chapter 3, we also have the continuity equation expressing the conservation of mass given as

$$\frac{d\varrho}{dt} + \varrho \, \frac{\partial v_i}{\partial x_i} = 0 \tag{10.35}$$

Finally, from equation (10.29) and the stress-rate definition of equation (10.19), we have the following elastic-plastic constitutive relations:

$$d_{ij} = \frac{1+\nu}{E} \, (\dot{t}_{ij} - \Omega_{ir}t_{rj} - \Omega_{jr}t_{ir}) - \frac{\nu}{E} \, \dot{t}_{kk} \, \delta_{ij}$$

$$+ \frac{\beta}{K} \left(t_{ij} - \frac{1}{3} \, t_{kk} \, \delta_{ij} \right) \tag{10.36}$$

where

$$\beta = \beta, \qquad J_2 = K^2$$
$$\beta = 0, \qquad J_2 < K^2$$

and where J_2 is given by

$$J_2 = \tfrac{1}{2}(t_{ij} - \tfrac{1}{3}t_{kk} \, \delta_{ij})(t_{ij} - \tfrac{1}{3}t_{kk} \, \delta_{ij}) \tag{10.37}$$

and K is related to the yield stress Y of the material in uniaxial loading by

$$K = Y/\sqrt{3} \tag{10.38}$$

Selected Reading

Malvern, L. E., *Introduction to the Mechanics of a Continuous Medium.* Prentice-Hall,
Englewood Cliffs, New Jersey, 1969. The equations of plasticity are discussed in
Chapter 6.
Hill, R., *The Mathematical Theory of Plasticity.* Clarendon Press, Oxford, England, 1950.
The classical Prandtl–Reuss equations of plasticity are developed in Chapter 2.

Exercises

1. By considering the strain energy function of Section 4.5 for an isotropic
   solid, show that the Maxwell–Mises yield condition can be regarded as
   a condition for yield when the distortional elastic energy (i.e., that
   part unaffected by changes in the mean stress) reaches a critical value.

2. Express the Levy–Mises constitutive relations in terms of principal
   stress and rate-of-strain values and show, by use of equation (10.32),
   that an increase in all the rates of strain leaves the stress unchanged.
   Contrast this result with that of a Kelvin–Voigt viscoelastic solid.

3. If the time scale required to achieve a given elastic-plastic deformation
   is doubled, will the stress as given by the Prandtl–Reuss equations remain unchanged?

11

Problems in Plasticity

In order to illustrate the theory of plasticity as developed in the previous chapter, we now proceed to consider a few selected problems requiring plasticity considerations for solution. Because of the inherently nonlinear form of the governing equations in plasticity theory, and the attendant mathematical complexities, these problems will be restricted to rather simple ones. Nevertheless they should serve to illustrate the general features involved in the plastic response of solids.

For sake of simplicity, we restrict our attention in all cases to static deformation or to quasistatic deformations where the plastic flow is assumed slow enough to allow the complete neglect of inertial effects.

11.1. Initial Yielding of a Thin-Walled Tube under Combined Tension–Torsion Loading

As a simple example of the use of the Maxwell–Mises yield condition, we consider the initial yielding of a thin-walled cylindrical tube of circular section subjected to combined tension and torsion loadings, as shown in Figure 11.1.

With polar coordinates r, θ, z as suggested, the nonzero stresses t_{zz} and $t_{\theta z}$ in the thin tube wall, which result from the tension and torsion loadings,

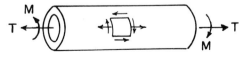

FIGURE 11.1.

may be assumed constant throughout the material. Choosing rectangular axes at any point with x_1, x_2, and x_3 lying in the radial, circumferential, and axial directions, we may thus take $t_{33} = t_{zz}$ and $t_{23} = t_{\theta z}$ as the nonzero stresses in the Maxwell–Mises condition given from equation (10.23) of Chapter 10 as

$$\tfrac{1}{2}(t_{ij} - \bar{t}\,\delta_{ij})(t_{ij} - \bar{t}\,\delta_{ij}) = K^2 \tag{11.1}$$

where $\bar{t} = t_{ii}/3$ and $K = Y/\sqrt{3}$, with Y denoting the yield stress in uniaxial loading. For this case, we find, on expanding, that

$$t_{33}^2 + 3t_{23}^2 = Y^2 \tag{11.2}$$

Adopting the notation

$$t_{33} = t_{zz} = \sigma, \qquad t_{23} = t_{\theta z} = \tau \tag{11.3}$$

we can write the above equation more simply as

$$\left(\frac{\sigma}{Y}\right)^2 + 3\left(\frac{\tau}{Y}\right)^2 = 1 \tag{11.4}$$

This equation provides general conditions on stresses σ and τ necessary for yielding. In particular, for the case of tensile loading only, the equation reduces, as it should, simply to the condition that $\sigma = Y$. Alternatively, in the case of torsion loading only, the equation reduces to the yield condition $\tau = Y/\sqrt{3}$, showing that the shear yield stress is reduced from the uniaxial yield stress by the factor $1/\sqrt{3}$. For intermediate cases where either the normal or shearing stress is specified, the equation may, of course,

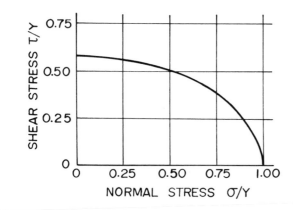

FIGURE 11.2.

then be used to determine the value of the unspecified stress necessary for yielding.

Figure 11.2 shows a graph of equation (11.4) for these cases of combined tension and torsion loadings. Experiments carried out on metal tubes subjected to these kinds of loadings have generally shown excellent agreement with this relation.

11.2. Initial Yielding of a Hollow Cylinder under Internal Pressure Loading

For a more complex illustration of the use of the Maxwell–Mises yield condition, we next consider the problem of determining the minimum internal pressure necessary to cause yielding in a hollow circular cylinder. From equation (5.127) of Chapter 5, we have the nonzero elastic stresses t_{rr} and $t_{\theta\theta}$ in the cylinder wall expressible as

$$t_{rr} = \frac{a^2 P}{b^2 - a^2}\left(1 - \frac{b^2}{r^2}\right) \tag{11.5}$$

$$t_{\theta\theta} = \frac{a^2 P}{b^2 - a^2}\left(1 + \frac{b^2}{r^2}\right) \tag{11.6}$$

where P denotes the internal pressure and where a and b denote the inside and outside radii as shown in Figure 11.3.

Due to the symmetry of the problem, the only remaining nonzero stress is the axial stress t_{zz} which, for plane strain, is expressible as

$$t_{zz} = \nu(t_{rr} + t_{\theta\theta}) = \frac{2\nu a^2 P}{b^2 - a^2} \tag{11.7}$$

Taking $t_{11} = t_{rr}$, $t_{22} = t_{\theta\theta}$, $t_{33} = t_{zz}$ and substituting into the Maxwell–Mises yield condition given by equation (11.1), we find after some reduction

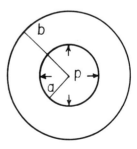

FIGURE 11.3.

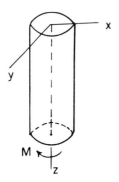

FIGURE 11.4.

that the pressure P required for initial yielding is expressible as

$$P = \frac{Y(b^2 - a^2)r^2}{a^2[(1 - 2\nu)^2r^4 + 3b^4]^{1/2}} \tag{11.8}$$

Since this relation provides the least value of P when $r = a$, we can thus expect initial yielding at the inside radius of the tube when the internal pressure first reaches the value given by

$$P = \frac{Y(b^2 - a^2)}{[(1 - 2\nu)^2a^4 + 3b^4]^{1/2}} \tag{11.9}$$

11.3. Twisting of a Circular Rod

We next consider the twisting of a circular rod by end torques and calculate the moment-twist relation that applies when elements of the rod are stressed to their yield point.

From equations (5.37) and (5.38) of Chapter 5, we have the elastic stresses t_{xz}, t_{yz} given for this case as

$$t_{xz} = -\mu\alpha y, \qquad t_{yz} = \mu\alpha x \tag{11.10}$$

where μ denotes the Lamé elastic shear constant, α denotes the angle of twist per unit length, and x and y denote position in the cross section, as shown in Figure 11.4.

If we adopt polar coordinates r, θ, z such that

$$r = (x^2 + y^2)^{1/2}$$
$$\theta = \tan^{-1}(y/x) \tag{11.11}$$
$$z = z$$

the above stress distribution can be written simply as

$$t_{\theta z} = \mu \alpha r \tag{11.12}$$

and the moment-twist relation is determined by integration as

$$M = 2\pi \int_0^R r t_{\theta z} r \, dr = \tfrac{1}{2}\pi\mu\alpha R^4 \tag{11.13}$$

This relation applies so long as the stress within the rod is entirely elastic. From the Maxwell–Mises yield condition, we can, however, easily see that yielding under the stress state defined by equation (11.12) will occur when the stress $t_{\theta z}$ satisfies

$$t_{\theta z} = Y/\sqrt{3} \tag{11.14}$$

where Y denotes the yield stress in uniaxial loading. Combining equations (11.12) and (11.14), we thus have yielding at the radius r of the rod when the angle of twist α satisfies

$$\alpha = Y/(\sqrt{3} \, \mu r) \tag{11.15}$$

Since, according to this relation, the angle of twist α is least when the radius r is greatest, yielding will first occur at the outside radius $r = R$ at a twist angle α_Y given by

$$\alpha_Y = Y/(\sqrt{3} \, \mu R) \tag{11.16}$$

and at a moment M_Y given from equation (11.13) as

$$M_Y = \tfrac{1}{2}\pi\mu R^4 \alpha_Y \tag{11.17}$$

As the angle of twist is increased past the above value of α_Y, the stress at smaller radial distances will likewise rise to the yield value given by equation (11.14), so that the rod will then consist of a central elastic region of radius r_Y together with a surrounding, outer plastic region, as shown in Figure 11.5. From equations (11.15) and (11.16), we have, in particular, the radius r_Y of the elastic region expressible as

$$r_Y = R\alpha_Y/\alpha \tag{11.18}$$

The stress distribution within the rod is thus given from equations (11.12) and (11.14) as

$$\begin{aligned} t_{\theta z} &= \mu\alpha r, & 0 \le r \le r_Y \\ t_{\theta z} &= Y/\sqrt{3}, & r_Y \le r \le R \end{aligned} \tag{11.19}$$

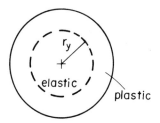

FIGURE 11.5.

For angles of twist greater than α_Y, the moment-twist relation is thus determined from

$$M = 2\pi \int_0^R r t_{\theta z} r \, dr \tag{11.20}$$

where $t_{\theta z}$ is given by equation (11.19). After some reduction, we find, in particular, for $\alpha \geq \alpha_Y$ that

$$M = \frac{4}{3} M_Y \left(1 - \frac{1}{4} \frac{\alpha_Y^3}{\alpha^3} \right) \tag{11.21}$$

where M_Y denotes the elastic moment associated with α_Y, as given by equation (11.17).

The moment-twist relation of the rod as determined from equations (11.13) and (11.21) for $\alpha \leq \alpha_Y$ and $\alpha \geq \alpha_Y$, respectively, is sketched in Figure 11.6.

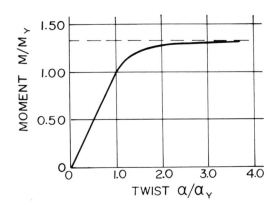

FIGURE 11.6.

11.4. Plastic Extension of a Cylindrical Bar under Simple Tension Loading

As a first illustration of the plastic constitutive relations developed in the previous chapter, we consider the gradual extension of a cylindrical bar which is loaded in simple tension. We choose, as indicated in Figure 11.7, spatial coordinates x_i such that axial position is denoted by $x_1 = x$ and transverse position by $x_2 = y$ and $x_3 = z$. The velocity components v_i for uniform extension are, accordingly, expressible as

$$v_1 = \dot{u}(x, t), \qquad v_2 = \dot{v}(y, t), \qquad v_3 = \dot{w}(z, t) \qquad (11.22)$$

where u, v, and w denote displacements in the x, y, and z directions and where dots denote usual material time derivatives.

Consider first the initial elastic loading of the bar. From equation (10.36) of Chapter 10, we have the constitutive relations for $t_{xx} < Y$ given as

$$d_{xx} = \frac{t_{xx}}{E}, \qquad d_{yy} = d_{zz} = -\frac{\nu}{E} t_{xx} \qquad (11.23)$$

where the rates of strain d_{xx}, d_{yy}, and d_{zz} are given by

$$d_{xx} = \frac{\partial \dot{u}}{\partial x}, \qquad d_{yy} = \frac{\partial \dot{v}}{\partial y}, \qquad d_{zz} = \frac{\partial \dot{w}}{\partial z} \qquad (11.24)$$

and where use has been made of the fact that the rate of rotation Ω_{ij} associated with the motion defined by equation (11.22) is zero. Integrating the first of equations (11.23) over the current length of the bar, from $x = 0$ to $x = l(t)$, we have

$$\dot{u}(l) - \dot{u}(0) = \frac{t_{xx}}{E} l \qquad (11.25)$$

FIGURE 11.7.

The difference $\dot{u}(l) - \dot{u}(0)$ represents, however, nothing more than the rate of increase of the length of bar, so that this last relation may be written more conveniently as

$$\frac{1}{l}\frac{dl}{dt} = \frac{1}{E}\frac{dt_{xx}}{dt} \tag{11.26}$$

On integrating this expression with respect to time and assuming $l = l_0$ and $t_{xx} = 0$ at time $t = 0$, we thus finally have

$$\log\frac{l}{l_0} = \frac{1}{E}t_{xx} \tag{11.27}$$

as the relation for describing the initial elastic response of the solid.

For small extensions, we also have $l/l_0 = 1 + e_{xx}$ and $\log(l/l_0) = e_{xx}$, where e_{xx} is the usual axial strain for small deformations, so that equation (11.27) is thus seen to reduce, as it should, to the uniaxial linear elastic relation of Chapter 4.

Consider next the case where the axial stress reaches the yield stress Y of the bar. From equation (10.36) of Chapter 10, we then have the constitutive relations given for $t_{xx} = Y$ as

$$d_{xx} = \frac{\beta}{K}(t_{xx} - \bar{t}) = \frac{2}{3}\frac{\beta}{K}Y$$

$$\tag{11.28}$$

$$d_{yy} = d_{zz} = \frac{\beta}{K}(-\bar{t}) = -\frac{1}{3}\frac{\beta}{K}Y$$

where

$$\bar{t} = t_{kk}/3 = Y/3 \quad \text{and} \quad K = Y/\sqrt{3}$$

It is easily seen that the relations of equation (11.28) together with the Maxwell–Mises yield condition provide the equation

$$d_{yy} = d_{zz} = -\tfrac{1}{2}d_{xx} \tag{11.29}$$

and the identity

$$d_{xx} = \tfrac{2}{3}(\beta/K)Y = d_{xx} \tag{11.30}$$

The first of these results can be seen from the continuity relation given by equation (10.35) of Chapter 10 to imply that the density, and hence the volume, of the material is unchanged during the plastic deformation. The second shows, moreover, that, with $t_{xx} = Y$, the axial rate of strain d_{xx} is determined only by displacement constraints existing on the ends of the bar and not at all by the constitutive relation of the material.

Finally, consider the case where after some amount of plastic deformation, the stress t_{xx} is reduced below the yield stress Y of the bar. The elastic constitutive relations for this situation are again given by equation (11.23). Integrating the first of these relations in the same manner as earlier and taking $l = l_1$ and $t_{xx} = Y$ at the instant when the elastic deformation begins, we find

$$\log \frac{l}{l_1} = \frac{1}{E}(t_{xx} - Y) \tag{11.31}$$

or, equivalently,

$$\log \frac{l}{l_0} - \log \frac{l_1}{l_0} = \frac{1}{E}(t_{xx} - Y) \tag{11.32}$$

where l_0 denotes the initial length of the bar. This result is valid for all axial stresses less than the yield stress of the material, regardless of the amount of prior plastic deformation and shows that the stress t_{xx} will be linearly related to the quantity $\log(l/l_0)$. If at some later time the stress is again raised to the yield stress, further plastic deformation will, of course, occur similar to that which followed the initial elastic deformation.

The above elastic–plastic response is illustrated in Figure 11.8. This response is generally found to approximate that of a number of hard structural metals and alloys when subjected to simple tension loading. It is worth noting, however, that there is a general tendency among metals, especially at large strains, to require increasingly larger stresses for continued plastic deformation. This phenomenon is referred to as *strain hardening* and is not included in the constitutive description of an elastic–perfectly plastic material as considered here.

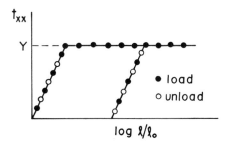

FIGURE 11.8.

11.5. Plane Strain Compression

The axial compression of a bar constrained against motion in one direction offers a simple yet instructive example of the effects of elasticity on the combined elastic–perfectly plastic response of a solid. The problem is illustrated in Figure 11.9, where spatial coordinates x_i have been chosen such that $x_1 = x$, $x_2 = y$, $x_3 = z$. The motion of the bar is assumed such that the velocity components v_i are expressible as

$$v_1 = \dot{u}(x, t), \qquad v_2 = \dot{v}(y, t), \qquad v_3 = \dot{w}(z, t) = 0 \tag{11.33}$$

where u, v, and w denote displacements in the x, y, and z directions and where dots denote usual material time derivatives.

From equation (10.36) of Chapter 10, the constitutive relations governing the initial elastic loading of the bar may be written for $J_2 < K^2$ as

$$d_{xx} = \frac{\partial \dot{u}}{\partial x} = \frac{1}{E}(\dot{t}_{xx} - \nu \dot{t}_{zz})$$

$$d_{yy} = \frac{\partial \dot{v}}{\partial y} = -\frac{\nu}{E}(\dot{t}_{xx} + \dot{t}_{zz}) \tag{11.34}$$

$$d_{zz} = \frac{\partial \dot{w}}{\partial z} = \frac{1}{E}(\dot{t}_{zz} - \nu \dot{t}_{xx}) = 0$$

where the stress in the y direction has been assumed zero and where account has been taken of the fact that the rate of rotation Ω_{ij} associated with the motion defined by equation (11.33) is zero.

The last of equations (11.34) gives immediately that

$$\dot{t}_{zz} = \nu \dot{t}_{xx} \tag{11.35}$$

and the first then becomes expressible as

$$\frac{\partial \dot{u}}{\partial x} = \frac{1 - \nu^2}{E} \dot{t}_{xx} \tag{11.36}$$

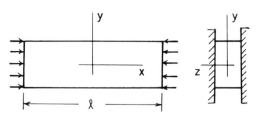

FIGURE 11.9.

If l denotes the instantaneous length of the bar, we also have, as in the previous section, that

$$\frac{\partial \dot{u}}{\partial x} = \frac{1}{l}\frac{dl}{dt} \tag{11.37}$$

Integrating these equations and taking $t_{zz} = 0$ and $l = l_0$ when $t_{xx} = 0$, we accordingly find

$$t_{zz} = \nu t_{xx} \tag{11.38}$$

$$\log\frac{l}{l_0} = \frac{1 - \nu^2}{E}t_{xx} \tag{11.39}$$

For small compressions $\log(l/l_0) = e_{xx}$, where e_{xx} denotes the usual small axial strain component, so that these relations are easily seen to be identical to those given by the elastic constitutive relations of Chapter 4.

Next consider the case of the initial yielding that occurs when t_{xx} reaches a value given by $J_2 = K^2$, that is, when

$$J_2 = \tfrac{1}{3}(t_{xx}^2 - t_{xx}t_{zz} + t_{zz}^2) = K^2 \tag{11.40}$$

where $K = Y/\sqrt{3}$, with Y denoting the yield stress in uniaxial loading. Using equation (11.38) with this relation, we find the initial yield values of t_{xx} and t_{zz} given by

$$t_{xx}^0 = \frac{-Y}{(1 - \nu + \nu^2)^{1/2}} \tag{11.41}$$

$$t_{zz}^0 = \frac{-\nu Y}{(1 - \nu + \nu^2)^{1/2}} \tag{11.42}$$

The corresponding length l_1 of the bar is also determined from equation (11.39) by the relation

$$\log\frac{l_1}{l_0} = \frac{1 - \nu^2}{E}t_{xx} \tag{11.43}$$

Finally, consider the plastic deformation that follows when t_{xx} is increased above the initial yield value given by equation (11.41). Once this happens, the values of the lateral stress t_{zz} and the longitudinal displacement are then no longer given by the elastic relations of equations (11.42) and (11.43), and they must be determined instead from the elastic–perfectly plastic constitutive relations given by equation (10.36) of Chapter 10 for

$J_2 = K^2$, namely

$$d_{xx} = \frac{\partial \dot{u}}{\partial x} = \frac{\beta}{3K}(2t_{xx} - t_{zz}) + \frac{1}{E}(\dot{t}_{xx} - \nu \dot{t}_{zz})$$

$$d_{yy} = \frac{\partial \dot{v}}{\partial y} = \frac{\beta}{3K}(-t_{xx} - t_{zz}) - \frac{\nu}{E}(\dot{t}_{xx} + \dot{t}_{zz}) \qquad (11.44)$$

$$d_{zz} = \frac{\partial \dot{w}}{\partial z} = \frac{\beta}{3K}(2t_{zz} - t_{xx}) + \frac{1}{E}(\dot{t}_{zz} - \nu \dot{t}_{xx}) = 0$$

From the last of these equations, we have

$$\frac{\beta}{K} = -\frac{3}{E}\frac{\dot{t}_{zz} - \nu \dot{t}_{xx}}{2t_{zz} - t_{xx}} \qquad (11.45)$$

so that the first may be written as

$$E\frac{\partial \dot{u}}{\partial x} = \frac{2t_{xx} - t_{zz}}{2t_{zz} - t_{xx}}(\nu \dot{t}_{xx} - \dot{t}_{zz}) + \dot{t}_{xx} - \nu \dot{t}_{zz} \qquad (11.46)$$

In addition, the condition $J_2 = K^2$ given by equation (11.40) provides the relation

$$t_{zz} = \tfrac{1}{2}t_{xx} + \tfrac{1}{2}(4Y^2 - 3t_{xx}^2)^{1/2} \qquad (11.47)$$

where the positive square root has been chosen for consistency with the initial yield conditions given by equations (11.41) and (11.42). Using this expression and its time derivative, we may therefore eliminate t_{zz} and $\dot{t}_{zz}$ from equation (11.46) to get

$$E\frac{\partial \dot{u}}{\partial x} = \left[\frac{9}{4}\frac{t_{xx}^2}{4Y^2 - 3t_{xx}^2} + \frac{3}{2}\frac{(2\nu - 1)t_{xx}}{(4Y^2 - 3t_{xx}^2)^{1/2}} + \frac{5 - 4\nu}{4}\right]\dot{t}_{xx} \qquad (11.48)$$

This equation may easily be integrated with the help of equation (11.37) to give

$$\log\frac{l}{l_0} = \frac{1 - 2\nu}{2E}t_{xx} + \frac{1 - 2\nu}{2E}(4Y^2 - 3t_{xx}^2)^{1/2}$$

$$+ \frac{\sqrt{3}}{4}\frac{Y}{E}\log\left(\frac{2Y + \sqrt{3}\,t_{xx}}{2Y - \sqrt{3}\,t_{xx}}\right) + C \qquad (11.49)$$

where the constant C is determined from the initial yield conditions of equations (11.41) and (11.43) as

$$C = \frac{1 + 2\nu - \nu^2}{2E} t_{xx}^0 - \frac{1 - 2\nu}{2E} [4Y^2 - 3(t_{xx}^0)^2]^{1/2}$$

$$- \frac{\sqrt{3}}{4} \frac{Y}{E} \log \frac{2Y + \sqrt{3}\, t_{xx}^0}{2Y - \sqrt{3}\, t_{xx}^0} \tag{11.50}$$

Equations (11.49) and (11.47) may be used to determine the stresses t_{xx} and t_{zz} in terms of $\log(l/l_0)$ when t_{xx} exceeds the yield value given by equation (11.41). For the initial stresses less than this yield value, equations (11.38) and (11.39) may similarly be used. The complete solutions for t_{xx}/Y and t_{zz}/Y are shown in Figure 11.10 as a function of $(E/Y) \log(l/l_0)$ for the typical case where $\nu = 0.3$.

Also shown in Figure 11.10 are the corresponding results which apply when the Levy–Mises rigid–perfectly plastic constitutive relations are used. These relations are given by equation (10.31) of Chapter 10 and here take the form

$$d_{xx} = \frac{\beta}{3K} (2t_{xx} - t_{zz})$$

$$d_{yy} = \frac{\beta}{3K} (-t_{xx} - t_{zz}) \tag{11.51}$$

$$d_{zz} = \frac{\beta}{3K} (2t_{zz} - t_{xx}) = 0$$

From the last of these, we have, in particular, that

$$t_{zz} = \tfrac{1}{2} t_{xx} \tag{11.52}$$

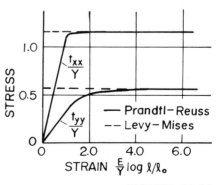

FIGURE 11.10.

and from the first two we find, with $K = Y/\sqrt{3}$, that

$$t_{xx} = 2Y/\sqrt{3} \tag{11.53}$$

It is worth noticing from the results of Figure 11.10 that the stresses given by the Prandtl–Reuss elastic–perfectly plastic constitutive relations very quickly approach those given by the Levy–Mises relations, which ignore the elasticity of the solid altogether.

11.6. Plane Strain Deformation of Rigid–Perfectly Plastic Solids

In discussing the plane-strain plastic deformation of solids that are free to flow in at least one direction, it is customary to neglect the elastic component of the deformation altogether and treat the problem using the simplified Levy–Mises rigid–perfectly plastic constitutive relations given by equation (10.31) of Chapter 10. As the example of the preceding section indicated, such a treatment can generally be expected to provide a good approximation to the more complex elastic–perfectly plastic treatment after only small plastic deformations of the order of the elastic yield deformation.

If we choose spatial coordinates x_i such that $x_1 = x$, $x_2 = y$, $x_3 = z$, the velocities v_i associated with plane strain deformations in the x–y plane may be written simply as

$$v_1 = \dot{u}(x, y, t), \qquad v_2 = \dot{v}(x, y, t), \qquad v_3 = \dot{w} = 0 \tag{11.54}$$

where u, v, and w denote displacements in the x, y, and z directions and where, as usual, dots denote material time derivatives.

With the above velocities, the Levy–Mises constitutive relations for perfectly plastic flow may next be written from equation (10.31) of Chapter 10 as

$$d_{xx} = \frac{\partial \dot{u}}{\partial x} = \frac{\beta}{3K} (2t_{xx} - t_{yy} - t_{zz}) \tag{11.55}$$

$$d_{yy} = \frac{\partial \dot{v}}{\partial y} = \frac{\beta}{3K} (2t_{yy} - t_{xx} - t_{zz}) \tag{11.56}$$

$$d_{zz} = \frac{\partial \dot{w}}{\partial z} = \frac{\beta}{3K} (2t_{zz} - t_{xx} - t_{yy}) = 0 \tag{11.57}$$

$$d_{xy} = \frac{1}{2}\left(\frac{\partial \dot{u}}{\partial y} + \frac{\partial \dot{v}}{\partial x}\right) = \frac{\beta}{K} t_{xy} \tag{11.58}$$

$$d_{yz} = \frac{1}{2}\left(\frac{\partial \dot{v}}{\partial z} + \frac{\partial \dot{w}}{\partial y}\right) = \frac{\beta}{K} t_{yz} = 0 \tag{11.59}$$

$$d_{xz} = \frac{1}{2}\left(\frac{\partial \dot{u}}{\partial z} + \frac{\partial \dot{w}}{\partial x}\right) = \frac{\beta}{K} t_{xz} = 0 \tag{11.60}$$

where, from equation (10.32) of Chapter 10, we have

$$\beta = [\tfrac{1}{2}(d_{xx}^2 + d_{yy}^2 + 2d_{xy}^2)]^{1/2}$$

and, as earlier,

$$K = Y/\sqrt{3}$$

with Y denoting, as usual, the yield stress in uniaxial loading.

From the third of these constitutive equations, we have immediately that

$$t_{zz} = \tfrac{1}{2}(t_{xx} + t_{yy}) \tag{11.61}$$

so that the first two may be written alternatively as

$$d_{xx} = \frac{\partial \dot{u}}{\partial x} = \frac{\beta}{2K}(t_{xx} - t_{yy}) \tag{11.62}$$

$$d_{yy} = \frac{\partial \dot{v}}{\partial y} = \frac{\beta}{2K}(t_{yy} - t_{xx}) \tag{11.63}$$

On adding these last two equations, we have immediately that

$$\frac{\partial \dot{u}}{\partial x} + \frac{\partial \dot{v}}{\partial y} = 0 \tag{11.64}$$

and, on combining equations (11.58) and (11.62), we find

$$(t_{xx} - t_{yy})\left(\frac{\partial \dot{u}}{\partial y} + \frac{\partial \dot{v}}{\partial x}\right) - 4t_{xy}\frac{\partial \dot{u}}{\partial x} = 0 \tag{11.65}$$

Equations (11.64) and (11.65) thus provide two relations for determining the velocities $\dot{u}$ and $\dot{v}$ when the stresses t_{xx}, t_{yy}, and t_{xy} are known.

Next consider the stress equilibrium equations and Maxwell–Mises yield condition for plane strain. From the constitutive relations of equations

(11.59) and (11.60), we may first observe that the stresses t_{yz} and t_{xz} are identically zero. Also, since the velocities $\dot{u}$ and $\dot{v}$ are independent of the out-of-plane coordinate z, it follows, in turn, from equations (11.62) and (11.63) and then (11.61) that the stresses t_{xx}, t_{yy}, and t_{zz} are independent of this coordinate. In these circumstances, the stress equilibrium equations become expressible in the absence of body forces simply as

$$\frac{\partial t_{xx}}{\partial x} + \frac{\partial t_{xy}}{\partial y} = 0 \tag{11.66}$$

$$\frac{\partial t_{xy}}{\partial x} + \frac{\partial t_{yy}}{\partial y} = 0 \tag{11.67}$$

In addition, the Maxwell–Mises yield condition becomes expressible simply as

$$(t_{xx} - t_{yy})^2 + 4t_{xy}^2 = 4K^2 \tag{11.68}$$

Equations (11.66)–(11.68) provide three relations connecting the three unknown stresses t_{xx}, t_{yy}, and t_{xy}. If suitable stress boundary conditions are given, these equations may thus be solved for the unknown stresses and the velocities $\dot{u}$ and $\dot{v}$ then determined from equations (11.64) and (11.65). Alternatively, if the boundary conditions on the problem involve the velocities, it is then necessary to consider simultaneously the five relations of equations (11.64)–(11.68).

11.7. Reduction of Plane Strain Equations

The above set of plane strain equations involves a system of partial differential equations whose solutions for particular problems are not easily obtained. By using the *theory of characteristics*, these equations may, however, be reduced to a set of ordinary differential equations which are generally more manageable.

To discuss this reduction, we first consider the stress relations given by equations (11.66)–(11.68). From the yield condition of equation (11.68), we have, in particular, that

$$t_{xy} = \pm \left[K^2 - \left(\frac{t_{xx} - t_{yy}}{2} \right)^2 \right]^{1/2} \tag{11.69}$$

so that, on substituting into the stress equilibrium conditions of equations

(11.66) and (11.67), we find these equations expressible as

$$\frac{\partial t_{xx}}{\partial x} - \gamma \frac{\partial t_{xx}}{\partial y} + \gamma \frac{\partial t_{yy}}{\partial y} = 0 \qquad (11.70)$$

$$\frac{\partial t_{yy}}{\partial y} + \gamma \frac{\partial t_{yy}}{\partial x} - \gamma \frac{\partial t_{xx}}{\partial x} = 0 \qquad (11.71)$$

where γ is given by

$$\gamma = \frac{t_{xx} - t_{yy}}{4t_{xy}} \qquad (11.72)$$

With these equations before us, we now ask whether there exists any curve C in the x–y plane along which the stresses t_{xx}, t_{yy}, and t_{xy} may be specified (consistent with the yield condition) without fixing their values in neighboring regions. Such curves, if they exist, are referred to as *characteristic curves*, or simply as *characteristics*, of equations (11.70) and (11.71).

To answer this question, let us first express the functions $t_{xx}(x, y)$ and $t_{yy}(x, y)$ at any instant as first-order Taylor series expansions about any point x_0, y_0 on the curve C of Figure 11.11. With Δx and Δy denoting small distances in the x and y directions from the point x_0, y_0 we have

$$t_{xx}(x, y) = t_{xx}(x_0, y_0) + \left(\frac{\partial t_{xx}}{\partial x}\right)_0 \Delta x + \left(\frac{\partial t_{xx}}{\partial y}\right)_0 \Delta y \qquad (11.73)$$

$$t_{yy}(x, y) = t_{yy}(x_0, y_0) + \left(\frac{\partial t_{yy}}{\partial x}\right)_0 \Delta x + \left(\frac{\partial t_{yy}}{\partial y}\right)_0 \Delta y \qquad (11.74)$$

where the zero subscript after the derivatives means that they are to be evaluated at x_0, y_0.

Now, if the stresses are assumed known everywhere along the curve C, the first terms on the right-hand side of these equations will be known.

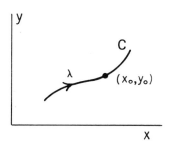

FIGURE 11.11.

Hence, if we can determine the derivatives $\partial t_{xx}/\partial_x$, etc., at the point x_0, y_0 using equations (11.70) and (11.71) and the known conditions along C, the above series expansions can be used to extend the solution for the stresses off the curve C to the point $x_0 + \Delta x$, $y_0 + \Delta y$. In this case, since x_0, y_0 denotes any point on C, this curve will therefore not be a characteristic curve. On the other hand, however, if we cannot determine any one of these derivatives uniquely, the curve C will, by definition, then be a characteristic curve of equation (11.70) and (11.71).

In order to attempt a solution for the four derivatives involved in equations (11.73) and (11.74), we may use the two relations given by equations (11.70) and (11.71) together with the two expressions that result from knowing the stresses—and hence their first derivatives—along the curve C, namely

$$\frac{dt_{xx}}{d\lambda} = \frac{\partial t_{xx}}{\partial x}\frac{dx}{d\lambda} + \frac{\partial t_{xx}}{\partial y}\frac{dy}{d\lambda} \tag{11.75}$$

$$\frac{dt_{yy}}{d\lambda} = \frac{\partial t_{yy}}{\partial x}\frac{dx}{d\lambda} + \frac{\partial t_{yy}}{\partial y}\frac{dy}{d\lambda} \tag{11.76}$$

where λ denotes distance along C and $x(\lambda)$ and $y(\lambda)$ denote coordinates of points on this curve.

Solving, for example, for $\partial t_{xx}/\partial x$ using the above four equations, we find, in particular, that

$$\frac{\partial t_{xx}}{\partial x} = \frac{1}{D}\begin{vmatrix} 0 & -\gamma & 0 & \gamma \\ 0 & 0 & \gamma & 1 \\ t'_{xx} & y' & 0 & 0 \\ t'_{yy} & 0 & x' & y' \end{vmatrix} \tag{11.77}$$

where D is given by

$$D = \begin{vmatrix} 1 & -\gamma & 0 & \gamma \\ -\gamma & 0 & \gamma & 1 \\ x' & y' & 0 & 0 \\ 0 & 0 & x' & y' \end{vmatrix} \tag{11.78}$$

and where primes denote derivatives with respect to λ.

Now, equation (11.77) will provide a unique solution for $\partial t_{xx}/\partial x$ at any point along C except when the denominator determinant vanishes, in which case the numerator determinant must also vanish if equation (11.77) is to have any solutions at all. Expanding the determinants for this case, we find, after some reduction and the use of equation (11.72), the following equations:

$$\left(\frac{dy}{dx}\right)^2 - \frac{4t_{xy}}{t_{xx} - t_{yy}} \frac{dy}{dx} - 1 = 0 \tag{11.79}$$

and

$$\left(1 - \gamma \frac{dy}{dx}\right) \frac{dt_{xx}}{d\lambda} + \gamma \frac{dy}{dx} \frac{dt_{yy}}{d\lambda} = 0 \tag{11.80}$$

Hence, if the slope of the curve C satisfies equation (11.79) at every point, then no unique solution for the derivative $\partial t_{xx}/\partial x$ is possible at points along C and the curve C is therefore a *stress characteristic curve* with the *stress characteristic relation* of equation (11.80) applying all along it.

In a similar way, we may also investigate the velocity relations given by equations (11.64) and (11.65) and determine equations describing their characteristic curves and characteristic relations. On carrying out this analysis, we find, in fact, that the equation defining a *velocity characteristic curve* is identical to equation (11.79) and that the *characteristic velocity relation* existing along it is expressible as

$$\frac{d\dot{u}}{d\lambda} + \frac{dy}{dx} \frac{d\dot{v}}{d\lambda} = 0 \tag{11.81}$$

With these results, we finally note that equation (11.79) can be solved for the slope dy/dx to get

$$\frac{dy}{dx} = \frac{2t_{xy}}{t_{xx} - t_{yy}} \pm \left[\left(\frac{2t_{xy}}{t_{xx} - t_{yy}}\right)^2 + 1\right]^{1/2} \tag{11.82}$$

This equation thus defines two *characteristic directions* at each point in the x–y plane where plastic deformation is occurring. In each of these directions, we thus have the stress and velocity relations of equations (11.80) and (11.81) applying, so that this resulting set of four ordinary differential equations, together with the yield condition, as given by equation (11.69), and equation (11.82) defining the characteristic directions, may be used in place of the original set of plane strain equations.

11.8. Slip-Line Theory

In using the above results to solve problems in plane strain, some simplification in the equations can be achieved by introducing certain geometric concepts. The resulting integration theory is then customarily referred to as *slip-line theory*.

We begin our discussion of this theory by introducing the angle ϕ_1 defined by

$$\tan \phi_1 = \frac{dy}{dx} \tag{11.83}$$

in which case, it may easily be seen that equation (11.79) defining all possible characteristics in the x–y plane may be written simply as

$$\tan 2\phi_1 = \frac{t_{yy} - t_{xx}}{2t_{xy}} \tag{11.84}$$

At any point in the x–y plane, this equation will provide two solutions $2\phi_1$ and $2\phi_1 + \pi$, so that, when the yield condition is satisfied, the angles to the two characteristic directions defined by equation (11.82) will be given simply by ϕ_1 and $\phi_1 + \pi/2$. At any point where plastic deformation is occurring, the two characteristic directions that exist there, which are tangent to the two characteristic curves passing through the point, will thus be perpendicular to one another.

If, as shown in Figure 11.12, we now choose a set of rectangular axes x', y' at any point P in the x–y plane such that the angle ϕ between the x' and x axes at that point is equal to either ϕ_1 or $\phi_1 + \pi/2$, as given by equation (11.84), the stresses t'_{xx}, t'_{yy}, and t'_{xy} referred to these axes may easily be seen from the tensor-transformation relation of Chapter 1 to be given by

$$t'_{xx} = \frac{t_{xx} + t_{yy}}{2} \pm \frac{t_{xx} - t_{yy}}{2} \cos 2\phi \pm t_{xy} \sin 2\phi$$

$$t'_{yy} = \frac{t_{xx} + t_{yy}}{2} \mp \frac{t_{xx} - t_{yy}}{2} \cos 2\phi \mp t_{xy} \sin 2\phi \tag{11.85}$$

$$t'_{xy} = \pm \frac{t_{yy} - t_{xx}}{2} \sin 2\phi \pm t_{xy} \cos 2\phi$$

where the upper sign corresponds to choosing $\phi = \phi_1$ and the lower sign to choosing $\phi = \phi_1 + \pi/2$.

From the last of the above transformation relations, it can easily be seen on forming the derivative of t'_{xy} with respect to ϕ and setting the result equal to zero that the directions of maximum and minimum shearing stresses at any point coincide with the characteristic directions defined by equation (11.84). For this reason, the characteristic curves are sometimes referred to as *shear lines* or *slip lines*.

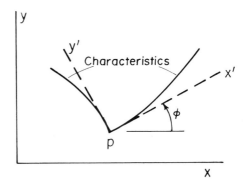

FIGURE 11.12.

Also, as a matter of convention, it is customary to choose the angle ϕ between the x and x' axes such that the shearing stress t'_{xy} is always positive, or, equivalently, such that the line of action of the algebraically larger principal stress lies in the first and third quadrants. When this is done, the x' and y' axes are then referred to as α and β axes, respectively. The characteristic curves to which these axes are tangent at a point are likewise customarily referred to as α and β lines.

Adopting the above convention for the choice of the angle, we have immediately from equations (11.84) and (11.85) and the yield condition, as expressed in the form of equation (11.69), that

$$t'_{xx} = t_{\alpha\alpha} = -p$$
$$t'_{yy} = t_{\beta\beta} = -p \qquad (11.86)$$
$$t'_{xy} = t_{\alpha\beta} = K$$

where p is given by

$$p = -\tfrac{1}{2}(t_{xx} + t_{yy}) \qquad (11.87)$$

These stresses are illustrated in Figure 11.13.

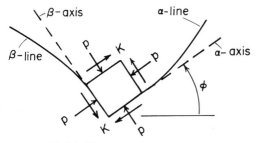

FIGURE 11.13.

If we now refer the above stresses back to the original x, y axes, we can easily find the results expressible as

$$t_{xx} = -p - K\sin 2\phi$$
$$t_{yy} = -p + K\sin 2\phi \qquad (11.88)$$
$$t_{xy} = K\cos 2\phi$$

Similarly, if $\dot{u}_\alpha$ and $\dot{u}_\beta$ denote velocity components in the α and β directions, we have also that

$$\dot{u} = \dot{u}_\alpha \cos\phi - \dot{u}_\beta \sin\phi$$
$$\dot{v} = \dot{u}_\alpha \sin\phi + \dot{u}_\beta \cos\phi \qquad (11.89)$$

On substituting the above stresses of equation (11.88) into the characteristic stress relation of equation (11.80), we find that, for characteristic α lines defined by

$$\frac{dy}{dx} = \tan\phi \qquad (11.90)$$

the result is expressible as

$$-\frac{dp}{d\lambda} - 2K\cos 2\phi \frac{d\phi}{d\lambda}(1 - 2\gamma \tan\phi) = 0 \qquad (11.91)$$

Noting further from equations (11.72) and (11.84) that γ may be written as

$$\gamma = -\tfrac{1}{2}\tan 2\phi \qquad (11.92)$$

we can express this relation simply as

$$-\frac{dp}{d\lambda} - 2K\frac{d\phi}{d\lambda} = 0 \qquad (11.93)$$

which applies on any α line.

Similarly, on taking

$$\frac{dy}{dx} = \tan(\phi + \pi/2) \qquad (11.94)$$

to define characteristic β lines, we also find

$$-\frac{dp}{d\lambda} + 2K\frac{d\phi}{d\lambda} = 0 \qquad (11.95)$$

on all β lines.

Integrating the above equations, we thus finally have the characteristic stress relations

$$p + 2K\phi = C_\alpha \quad \text{on} \quad \alpha \text{ lines}$$
$$p - 2K\phi = C_\beta \quad \text{on} \quad \beta \text{ lines}$$

(11.96)

where, for fixed α and β lines, C_α and C_β denote constants, which generally vary from line to line.

On substituting the velocity relations given by equation (11.89) into the characteristic velocity relation of equation (11.81), we can also find in a like manner that

$$\frac{d\dot{u}_\alpha}{d\lambda} - \dot{u}_\beta \frac{d\phi}{d\lambda} = 0 \quad \text{on} \quad \alpha \text{ lines}$$
$$\frac{d\dot{u}_\beta}{d\lambda} + \dot{u}_\alpha \frac{d\phi}{d\lambda} = 0 \quad \text{on} \quad \beta \text{ lines}$$

(11.97)

11.9. Numerical Solutions Using Slip-Line Theory

The above slip-line theory may be used to determine the stresses and velocities within any plastic region where suitable boundary conditions are prescribed. We discuss below the procedures to be followed for the three boundary value problems most frequently encountered.

First Boundary Value Problem. We consider the problem of determining the stresses and velocities in the plastic region ABC adjacent to a curve AB along which stresses (consistent with the yield condition) and velocities are known (Figure 11.14). We construct an orthogonal net

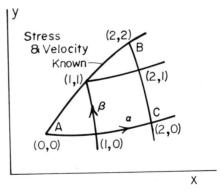

FIGURE 11.14.

of α and β lines away from the boundary and label the points as indicated, with (m, n) denoting the intersection of the mth β line with the nth α line. We assume also that the boundary AB is not itself a slip line. As a specific example, we now determine the stress, velocity, and x and y coordinates of the point $(2,1)$ from a knowledge of the boundary conditions along AB.

From equation (11.96), we have, first of all, the relations

$$p + 2K\phi = C_\alpha \qquad \text{on} \quad \alpha \text{ lines}$$
$$p - 2K\phi = C_\beta \qquad \text{on} \quad \beta \text{ lines} \tag{11.98}$$

Since the stress is known at point $(1,1)$ on the boundary, we can determine the constant C_α for the α line containing points $(1,1)$ and $(2,1)$ as

$$C_\alpha = p_{1,1} + 2K\phi_{1,1}$$

Similarly, on the β line containing the points $(2,2)$ and $(2,1)$, we also have

$$C_\beta = p_{2,2} - 2K\phi_{2,2}$$

Hence, on applying equations (11.98) and solving for $p_{2,1}$ and $\phi_{2,1}$, we obtain

$$p_{2,1} = \tfrac{1}{2}(p_{1,1} + p_{2,2}) + K(\phi_{1,1} - \phi_{2,2})$$
$$\phi_{2,1} = \tfrac{1}{2}(p_{1,1} - p_{2,2}) + K(\phi_{1,1} + \phi_{2,2}) \tag{11.99}$$

With these values, the stress components are then determined from equations (11.88) as

$$(t_{xx})_{2,1} = -p_{2,1} - K\sin 2\phi_{2,1}$$
$$(t_{yy})_{2,1} = -p_{2,1} + K\sin 2\phi_{2,1} \tag{11.100}$$
$$(t_{xy})_{2,1} = K\cos 2\phi_{2,1}$$

Next we locate the x and y coordinates of point $(2,1)$. As an approximation, we write the slip-line equation,

$$\frac{dy}{dx} = \tan\phi$$

for the α line containing points $(1,1)$ and $(2,1)$ in the form

$$y_{2,1} - y_{1,1} = (x_{2,1} - x_{1,1})\tan[\tfrac{1}{2}(\phi_{1,1} + \phi_{2,1})] \tag{11.101}$$

Similarly, we write the slip-line equation

$$\frac{dy}{dx} = \tan\left(\phi + \frac{\pi}{2}\right) = -\cot\phi$$

for the β line containing points (2,2) and (2,1) as

$$y_{2,1} - y_{2,2} = -(x_{2,1} - x_{2,2}) \cot[\tfrac{1}{2}(\phi_{2,2} + \phi_{2,1})] \qquad (11.102)$$

Knowing the coordinates of the boundary points (1,1) and (2,2), we can thus solve the above equations simultaneously for the coordinates of point (2,1).

Finally, consider the determination of the velocity components at point (2,1). The boundary conditions on AB provide the velocity components $\dot{u}_\alpha$, $\dot{u}_\beta$ at the points (1,1) and (2,2) of Figure 11.14. The velocity components at the point (2,1) may therefore be expressed in terms of these known velocity components by using a finite-difference approximation to the slip-line relations of equation (11.97) in the form

$$(\dot{u}_\alpha)_{2,1} - (\dot{u}_\alpha)_{1,1} = \tfrac{1}{2}[(\dot{u}_\beta)_{2,1} + (\dot{u}_\beta)_{1,1}](\phi_{2,1} - \phi_{1,1})$$
$$(\dot{u}_\beta)_{2,1} - (\dot{u}_\beta)_{2,2} = -\tfrac{1}{2}[(\dot{u}_\alpha)_{2,1} + (\dot{u}_\alpha)_{2,2}](\phi_{2,1} - \phi_{2,2}) \qquad (11.103)$$

On solving these equations simultaneously, we can thus determine the velocity components $(\dot{u}_\alpha)_{2,1}$ and $(\dot{u}_\beta)_{2,1}$.

It is easily seen that the above procedure can similarly be used to determine the stresses and velocities at all points within the region ABC of Figure 11.14. This region, which is bounded by the curve AB and by α and β lines from A and B, respectively, is referred to as the *domain of influence* of the boundary data along AB.

Second Boundary Value Problem. The above method of solution involved the assumption of stress and velocity data along a curve which was not itself a slip line. A similar solution procedure may also be employed when stress and normal-velocity-component data are specified along two intersecting slip lines such as AB and AD of Figure 11.15.

Paralleling the above procedure, we first construct an orthogonal net of slip lines away from the known slip lines AB and AD and label the points as indicated. The domain of influence for the boundary data on AB and AD in this case is the region $ABCD$. To calculate the stress at the point (1,1), for example, we consider the α line containing the point (0,1) and the β line containing the point (1,0) and write equations (11.96) in the form

$$p_{1,1} + 2K\phi_{2,2} = p_{0,1} + 2K\phi_{0,1}$$
$$p_{1,1} - 2K\phi_{1,1} = p_{1,0} - 2K\phi_{1,0} \qquad (11.104)$$

where the terms on the right-hand side are all known because of the known

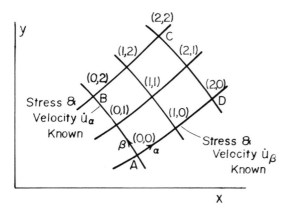

FIGURE 11.15.

stress values on the boundary slip lines AB and AD. These equations may therefore be solved for $p_{1,1}$ and $\phi_{1,1}$ and the stress components at point $(1,1)$ then determined, as in the previous problem, from equation (11.88) in the form

$$(t_{xx})_{1,1} = -p_{1,1} - K \sin 2\phi_{1,1}$$
$$(t_{yy})_{1,1} = -p_{1,1} + K \sin 2\phi_{1,1} \qquad (11.105)$$
$$(t_{xy})_{1,1} = K \cos 2\phi_{1,1}$$

The x and y coordinates of point $(1,1)$ are likewise determined as in the previous example by simultaneous solution of the equations

$$y_{1,1} - y_{0,1} = (y_{1,1} - x_{0,1}) \tan[\tfrac{1}{2}(\phi_{1,1} + \phi_{0,1})]$$
$$y_{1,1} - y_{0,1} = -(x_{1,1} - x_{1,0}) \cot[\tfrac{1}{2}(\phi_{1,1} + \phi_{1,0})] \qquad (11.106)$$

Finally, the velocity components at the point $(1,1)$ are determined, as in the previous problem, by simultaneous solution of the equations

$$(\dot{u}_\alpha)_{1,1} - (\dot{u}_\alpha)_{0,1} = \tfrac{1}{2}[(\dot{u}_\beta)_{1,1} + (\dot{u}_\beta)_{0,1}](\phi_{1,1} - \phi_{0,1})$$
$$(\dot{u}_\beta)_{1,1} - (\dot{u}_\beta)_{1,0} = -\tfrac{1}{2}[(\dot{u}_\alpha)_{1,1} + (\dot{u}_\alpha)_{1,0}](\phi_{1,1} - \phi_{1,0}) \qquad (11.107)$$

It is worth noting in these equations that it is only the $\dot{u}_\beta$ component of velocity that is needed on the bounding α slip line AD and only the $\dot{u}_\alpha$ component that is needed on the bounding β slip line AB.

Third Boundary Value Problem. If the stress and the normal velocity component are known along a single slip line and the value of ϕ and a velocity relation $f(\dot{u}_\alpha, \dot{u}_\beta) = 0$ are known along an intersecting curve, we can also determine the solution in the adjacent region using procedures similar to those discussed above. Consider, for example, Figure 11.16, where the segment AB denotes a curve along which ϕ and the velocity relation are known and where AC denotes an α line along which the stress and the normal velocity component $\dot{u}_\beta$ are known. We construct the orthogonal net of slip lines as shown within the domain of influence ABC and determine, as an example, the stress and velocity at the point (2,1). Considering the α and β lines containing points (1,1) and (2,0), respectively, we can write equation (11.96) in the form

$$p_{2,1} + 2K\phi_{2,1} = p_{1,1} + 2K\phi_{1,1}$$
$$p_{2,1} - 2K\phi_{2,1} = p_{2,0} - 2K\phi_{2,0} \tag{11.108}$$

where all but $p_{1,1}$ on the right-hand side of these equations is known from the boundary data. Considering next the β line containing point (1,0), we also have from equation (11.96) that

$$p_{1,1} - K\phi_{1,1} = p_{1,0} - 2K\phi_{1,0} \tag{11.109}$$

where all quantities but $p_{1,1}$ are known. Hence, using this relation to solve for $p_{1,1}$, we may then return to equations (11.108) and determine $p_{2,1}$ and $\phi_{2,1}$. Having these values, we can determine the stress components from equation (11.88) as in the previous examples. The location of the x and y coordinates can likewise be determined in the same way as in the earlier examples.

Finally, the velocity at the point (2,1) can be determined as in the first example provided the velocity components are first established at points

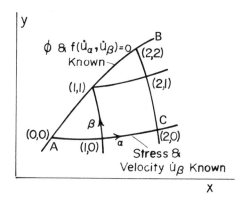

FIGURE 11.16.

along the boundary AB. These components are easily established using the velocity relation along β lines and the functional relation between the velocity components on AB. At the point $(1,1)$, we have, for example,

$$(\dot{u}_\beta)_{1,1} - (\dot{u}_\beta)_{1,0} = -\tfrac{1}{2}[(\dot{u}_\alpha)_{1,1} + (\dot{u}_\alpha)_{1,0}](\phi_{1,1} - \phi_{1,0})$$

$$f[(\dot{u}_\alpha)_{1,1}, (\dot{u}_\beta)_{1,1}] = 0 \tag{11.110}$$

which may be solved for $(\dot{u}_\alpha)_{1,1}$ and $(\dot{u}_\beta)_{1,1}$.

Special Cases. The above relations simplify considerably when straight slip lines are involved. Moreover, if one slip line in a family of α or β lines is straight, it is easily shown that all slip lines in that family must be straight. To see this, consider the intersecting α and β lines of Figure 11.17. By applying the slip-line relations of equation (11.96) between each of the four points A, B, C, and D, we find, after some reduction, that the following relations must apply:

$$\phi_B - \phi_A = \phi_C - \phi_D$$

$$\phi_D - \phi_A = \phi_C - \phi_B \tag{11.111}$$

This is *Hencky's theorem.* If, for example, the α line AD is straight, then $\phi_C - \phi_A = 0$ and equation (11.111) shows that the corresponding α line BC must also be straight. From these results, we thus see immediately that a slip-line net containing straight lines consists of either two orthogonal families of parallel, straight slip lines or one family of straight slip lines and a corresponding family of orthogonal lines.

Consider first the case of two orthogonal families of parallel, straight slip lines (Figure 11.18). Since ϕ is constant throughout the region containing these lines, we may immediately apply equation (11.104) to find that p is

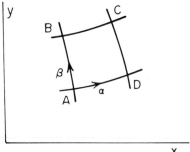

FIGURE 11.17.

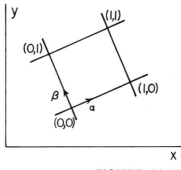

FIGURE 11.18.

also constant throughout this region. In this case, equation (11.105) then shows that the state of stress throughout this region is uniform. Similarly, equation (11.107) shows that the velocity component $\dot{u}_\alpha$ is constant along any α line and that $\dot{u}_\beta$ is constant along any β line.

Next consider the case in which the slip lines of one family are straight and the slip lines of the other family are orthogonal to these. Suppose further that all the straight slip lines meet at a point A. This situation is illustrated in Figure 11.19 for the case where the β lines are straight and the α lines are orthogonal to these. Such a net of slip lines is referred to as a *simple fan*. Since ϕ is constant along any β line, the second of equations (11.104) shows that p will be constant along any β line. Thus, both ϕ and p will be constant along the β line AB of Figure 11.19. In this case, the first of equations (11.104) then shows that p must be given everywhere in the simple-fan region by

$$p = c - 2K\phi \qquad (11.112)$$

where c is a constant denoting the constant value of $p + 2K\phi$ along the

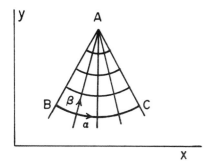

FIGURE 11.19.

β line *AB*. At the intersection point *A*, *p* is, of course, undefined, so that point *A* is a *singular point* where the stress is itself undefined. With regard to the velocity components, equation (11.107) shows simply that the component $\dot{u}_\beta$ is constant along a straight β line.

11.10. Wedge Penetration in a Rigid-Plastic Material

In the previous section, we considered the calculation of stresses and velocities in the interior of a plastic region for the case where the stress boundary conditions were sufficient to allow the construction of the slip lines without recourse to the velocity boundary conditions. Problems of this kind are customarily referred to as *statically determinate problems*. In many practical problems, however, the stress boundary conditions are not sufficiently specified to allow the complete determination of the stresses and slip lines so that both stress and velocity boundary conditions must be considered. The simplest approach to problems of this kind is to make tentative assumptions about the slip lines and subsequently examine whether all stress and velocity boundary conditions are satisfied. We illustrate this method of solution by considering the problem of penetration of a semi-infinite block of rigid-plastic material by a lubricated rigid wedge of total angle 2θ, as shown in Figure 11.20.

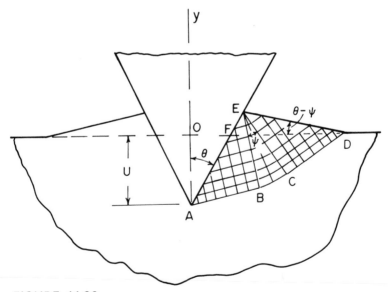

FIGURE 11.20.

Since the wedge is assumed frictionless, the wedge–material interface *AFE* shown in Figure 11.20 will be free of shearing stress. From equation (11.84), it can thus be seen that the slip lines must meet the interface *AFE* at 45°. Similarly, the slip lines must also meet the free surface *ED* of the material at 45°. In addition, since the free surface *DE* will not necessarily meet the wedge orthogonally, point *E* must be a singular point where the stresses are undefined. With these facts before us, we may therefore assume tentatively two uniform and properly inclined slip-line fields adjacent to the surfaces *AFE* and *ED*, with a simple fan region *EBC* connecting them, as shown in Figure 11.20.

In these circumstances, we have from the geometry of the triangle *FDE* that the angle *FDE* must equal $\theta - \psi$, where ψ is the angular span of the slip-line fan *EBC*. Also from the geometry of Figure 11.20, we have

$$U = AE \cos \theta - ED \sin(\theta - \psi) \tag{11.113}$$

where *U* denotes the depth of penetration at any instant. Since

$$AE = ED = h \tag{11.114}$$

this may be written alternatively as

$$U = h[\cos \theta - \sin(\theta - \psi)] \tag{11.115}$$

In addition, since we are dealing with a rigid-plastic material where the density and, hence, the volume of the material must remain constant, we have the further requirement that the volume of the material displaced by the wedge must equal the volume of the material above the original surface of the material. This condition yields from Figure 11.20 the equation

$$\tfrac{1}{2}U^2 \tan \theta = \tfrac{1}{2}(DF)(ED) \sin(\theta - \psi) \tag{11.116}$$

where *DF* is given from the geometry of the triangle *FED* as

$$DF = \frac{\cos \psi}{\cos \theta} ED = \frac{\cos \psi}{\cos \theta} h \tag{11.117}$$

On combining the above two equations, we thus have

$$U^2 \sin \theta = h^2 \cos \psi \sin(\theta - \psi) \tag{11.118}$$

If we next eliminate the ratio U/h between equations (11.115) and (11.118), we find

$$(\sin \theta)[\cos \theta - \sin(\theta - \psi)]^2 = \cos \psi \sin(\theta - \psi) \qquad (11.119)$$

which accordingly determines the fan angle ψ when the wedge angle θ is given.

With the angle ψ determined by the above equation, we have the assumed slip-line field of Figure 11.20 completely specified. It remains to examine the stress and velocity boundary conditions. Consider first the stress conditions. Since ED is free from normal and shearing stresses, we may easily establish from the yield condition that the stress component parallel to ED is equal to $-2K$. Thus, the stress is completely specified along the surface ED. In these circumstances, it is easily found that the α lines meeting this surface must be directed upward and to the right, as shown in Figure 11.20, in order to satisfy the convention established earlier in connection with equation (11.86). The stress may therefore be determined everywhere within the plastically deforming region by methods of the first boundary value problem discussed in the last section.

In particular, it is easily established that the mean pressure p and the angle ϕ between the α lines and the horizontal x axis must have the following values on this surface:

$$p = K, \qquad \phi = \tfrac{1}{4}\pi + (\psi - \theta) \qquad (11.120)$$

From equation (11.96), we thus have on all α lines that

$$p + 2K\phi = K + 2K(\tfrac{1}{4}\pi + \psi - \theta) \qquad (11.121)$$

On choosing any point in the plastically deforming region and determining the angle ϕ between the α line and the horizontal x axis, we may use this equation to find the mean stress p and then use equations (11.88) to determine the stresses t_{xx}, t_{yy}, and t_{xy}. The stress boundary conditions are thus all satisfied and the stresses are determined everywhere within the plastic region formed by the wedge penetration.

Next consider the velocity boundary conditions. Since $ABCD$ represents the rigid-plastic boundary, we have $\dot{u}_\beta = 0$ everywhere along this boundary. In addition, since the wedge penetration is assumed frictionless, the velocity imparted to the wedge–material interface AFE will be normal to it and of magnitude $\dot{U} \sin \theta$, where $\dot{U}$ denotes the velocity of the wedge penetration. The velocity everywhere in the region ABE is thus completely determined by the methods of the third boundary value problem discussed in the previous section. The velocity in the region *BED* may likewise be determined by the methods of the second boundary value problem.

We find, in particular, that the velocity component $\dot{u}_\beta$ must be zero on all β lines and that the velocity component $\dot{u}_\alpha$ must be given on all α lines by

$$\dot{u}_\alpha = \sqrt{2}\ \dot{U} \sin \theta \qquad (11.122)$$

These results satisfy all velocity boundary conditions and show that, as the wedge penetrates into the material, it pushes the displaced material upward and outward along the α slip lines with uniform speed $\dot{u}_\alpha$.

Selected Reading

Hill, R., *The Mathematical Theory of Plasticity*. Clarendon Press, Oxford, England, 1950. Chapters 6–9 contain an extensive discussion of problems involving plane-strain slip-line theory.

Prager, W., and P. G. Hodge, *Theory of Perfectly Plastic Solids*. John Wiley and Sons, New York, 1951. This book gives a readable account of the theory of plasticity. Chapters 2 and 3 deal with bending and torsion of bars.

Hill, R., E. H. Lee, and S. J. Tupper, "The Theory of Wedge Indentation of Ductile Materials," *Proceedings of the Royal Society (A)* **188**, 273–289 (1947). This paper gives the wedge-penetration solution discussed here.

Bell, J. F., "The Experimental Foundations of Solid Mechanics," in *Encyclopedia of Physics*, Vol. VIa/1, Springer-Verlag, Berlin, 1973. A monumental treatise on experimental solid mechanics, including a detailed history of the subject and an extensive discussion of plasticity.

Exercises

1. Show that the yielding of the thin-walled tube of Section 11.1 is governed by the equation

$$\left(\frac{\sigma}{Y}\right)^2 + 4\left(\frac{\tau}{Y}\right)^2 = 1$$

when *Tresca's maximum shear stress condition* is assumed in place of the Maxwell–Mises condition.

2. Consider a wide block being compressed in plane-strain between *rough* rigid plates such that the following boundary conditions apply:

$$t_{xy} = K, \qquad y = +h$$
$$t_{xy} = -K, \qquad y = -h$$

where y is measured from the mid-surface. Assuming a solution such that t_{xy} depends on y only, show that, for rigid–perfectly plastic deformation, the stresses are given by

$$t_{xx} = -P + \frac{Kx}{h} + \frac{2K}{h}(h^2 - y^2)^{1/2}$$

$$t_{yy} = -P + \frac{Kx}{h}$$

$$t_{xy} = -K\frac{y}{h}$$

where P denotes an arbitrary constant determined by edge conditions. This is *Prandtl's cycloid solution*.

3. Reconsider the wedge penetration solution of Section 11.10 for the case where the fan *ECB* extends all the way over to the wedge surface *AFE* and show that this assumed solution solves the problem of *rough* wedge penetration.

Appendix A

Similitude and Scale Modeling in Solid Mechanics

The use of dimensional analysis in problems in solid mechanics leads to general results which are frequently of considerable value in practical applications. The method is based on the observation that any equation describing physical phenomena must be invariant under change in the physical units of the variables involved and, hence, must be expressible in terms of dimensionless ratios of these variables.

Elasticity Example. To illustrate the method, let us consider the simple problem of an elastic body of arbitrary but specified shape which is acted on by forces F as shown in Figure A.1.

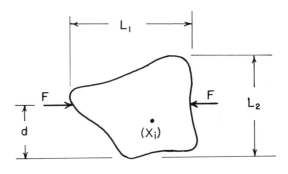

FIGURE A.1.

259

From our earlier study of elasticity, we may expect that the stress t_{ij} existing at any point X_i in the solid must be expressible as

$$t_{ij} = \phi_{ij}(F, L_1, L_2, d, X_k, E, \nu) \tag{A.1}$$

where ϕ_{ij} denotes an unknown function, E and ν denote Young's modulus and Poisson's ratio of the material (assumed isotropic), and the remaining variables are as defined in Figure A.1.

Without loss of generality, we may obviously express equation (A.1) in the form

$$\frac{t_{ij}}{E} = f_{ij}\left(\frac{F}{EL_1^2}, \frac{L_2}{L_1}, \frac{d}{L_1}, \frac{X_k}{L_1}, \nu, E, L_1\right) \tag{A.2}$$

where f_{ij} denotes an unknown function and where we have arbitrarily used the variables E and L_1 to divide out the force and length dimensions of the remaining variables. Mathematically, equation (A.2) is, of course, equivalent to equation (A.1) and one can be obtained from the other by simple choice of the unspecified functions. However, since the only remaining dimensional variables in equation (A.2) are E and L_1 and since there is no way that they can be combined to yield dimensionless quantities, these variables cannot appear separately if the equation is to be invariant under change in dimensions. Under this condition, equation (A.2) thus reduces to

$$\frac{t_{ij}}{E} = f_{ij}\left(\frac{F}{EL_1^2}, \frac{L_2}{L_1}, \frac{d}{L_1}, \frac{X_k}{L_1}, \nu\right) \tag{A.3}$$

If instead of the stress t_{ij} we had considered the displacement u_i at place X_i in the body, we would have found similarly the reduced equation

$$\frac{u_i}{L_1} = g_i\left(\frac{F}{EL_1^2}, \frac{L_2}{L_1}, \frac{d}{L_1}, \frac{X_k}{L_1}, \nu\right) \tag{A.4}$$

where g_i denotes an unknown function.

From the above equations, we easily see that if the arbitrarily chosen lengths L_1 and L_2 of the body (and, hence, all dimensions of the body) and the distance d are all changed by the same amount α, and the force is changed by the amount α^2, the stress in the body will be unchanged at the corresponding location αX_i and the associated displacement will change by the amount α, provided E and ν do not change; that is, provided the same material is used in both cases.

The above statement represents, of course, a simple example of a *scaling law* for elastic solids. It is worth noting that the law holds not only for linear elastic solids but also for any solid characterized by material constants having, at most, dimensions of stress. Thus, for example, the law also holds for non linear elastic solids as well as elastic-plastic solids. Such laws are clearly of value in experimental work as well as in applying theoretical and numerical results to particular cases.

Buckingham's Pi Theorem. If we compare equation (A.3) with equation (A.1), we see that dimensional reasoning allowed the original set of eight variables in the equation to be reduced by the number of independent dimensions involved, in this case two. This result when generalized leads to *Buckingham's Pi Theorem* in the form:

If m variables having n independent units appear in a physical equation, these variables may be replaced by $m - n$ independent dimensionless ratios of the original variables.

Viscoelastic Example. As a second example, consider the problem of a heavy rigid body of mass M impacting a mat of viscoelastic material with striking velocity V_0, as shown in Figure A.2.

The maximum acceleration a of the body when in contact with the material can be expected to have the following functional dependence:

$$a = a(M, V_0, \varrho, D_1, D_2, \lambda, \mu, \tilde{\lambda}, \tilde{\mu}, \lambda^*, \mu^*) \qquad (A.5)$$

where ϱ denotes the density of the material, D_1 and D_2 denote typical dimensions, and $\lambda, \mu, \ldots, \mu^*$ denote viscoelastic material constants. Using Buckingham's Pi Theorem, this relation may be written as

$$\frac{aD_1}{V_0^2} = f\left(\frac{MV_0^2}{D_1^3\lambda}, \frac{\varrho D_1^3}{M}, \frac{\tilde{\lambda}V_0}{\lambda D_1}, \frac{D_2}{D_1}, \frac{\lambda^*V_0}{D_1}, \frac{\lambda}{\mu}, \frac{\tilde{\lambda}}{\tilde{\mu}}, \frac{\lambda^*}{\mu^*}\right) \qquad (A.6)$$

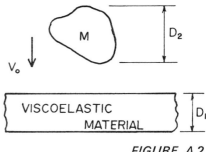

FIGURE A.2.

If we restrict attention always to the same material, the first two independent dimensionless variables require that the striking velocity be held constant during change in scale, while the third and fifth require that it vary directly with the length D_1. Thus, no scaling law exists, in general, for this problem. On the other hand, if we further restrict attention to the case of a relatively massive body striking a relatively light material, the second term in the above equation can be ignored and the equation then yields the following scaling law: If the linear dimensions of the system, the striking velocity, and the mass of the body are all scaled by the factor α, the acceleration will likewise scale by the factor α, provided the same material is employed in all cases. This is *Soper's scaling law* for cushioning mechanics. We note that the law obviously holds, not only for linear viscoelastic materials, but also for any material possessing one or more material constants having the dimensions of stress and one or more constants having the dimensions of time or combinations of time and stress.

Selected Reading

Langhaar, H. L., *Dimensional Analysis and Theory of Models*. John Wiley and Sons, New York, 1951. A readable introduction to the subject.

Soper, W. G., "Scale Modeling," *International Science and Technology*, February 1967, pp. 60–70. An excellent discussion of the applications of scale modeling in solid mechanics.

Soper, W. G., "Dynamic Modeling with Similar Materials," in *Use of Models and Scaling in Shock and Vibration*, pp. 51–56. American Society of Mechanical Engineers, New York, 1963. Excellent examples of scale modeling in solids.

Appendix B

Introduction to Numerical Methods in Solid Mechanics

The equations of solid mechanics are generally too complex to allow solutions to practical problems having detailed geometry and loading conditions. Fortunately, however, numerical methods of analysis are available for use when solutions to such problems are required. In dealing with equilibrium problems in solid mechanics, the two main methods of analysis are the *finite difference method* and the *finite element method*. We briefly indicate below the chief features of these methods.

B.1. Finite Difference Method

In applying this method, the continuum solid is approximated by a finite number of interior and boundary points and the governing differential equation and boundary conditions are replaced by approximate difference equations which are required to hold at each interior and boundary point, respectively. The method leads to a system of algebraic equations which can be solved to obtain desired values at the discrete points of the body.

One-Dimensional Example. We consider the simple problem of a rectangular, simply supported beam of flexural rigidity EI having a uniform loading p, as shown in Figure B.1. From the strength-of-material formulation of Chapter 5, we have the governing differential equation and boundary

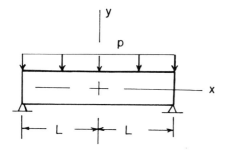

FIGURE B.1.

conditions expressible in terms of the deflection v as

$$EI \frac{d^4v}{dx^4} = -p$$

$$v = \frac{d^2v}{dx^2} = 0 \qquad \text{at} \quad x = -L, L$$

(B.1)

Choosing dimensionless variables

$$\bar{v} = v \frac{EI}{pL^4}, \qquad \bar{x} = \frac{x}{L}$$

(B.2)

these equations become

$$\frac{d^4\bar{v}}{d\bar{x}^4} = -1$$

$$\bar{v} = \frac{d^2\bar{v}}{d\bar{x}^2} = 0 \qquad \text{at} \quad \bar{x} = -1, 1$$

(B.3)

We now consider three interior points and the two boundary points, as shown in Figure B.2, and attempt to obtain a finite difference solution for the deflection at the interior points.

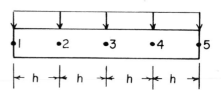

FIGURE B.2.

If we expand the deflection $\bar{v}$ in a Taylor series about some point, say the ith point, we may express the deflections $\bar{v}_{i-2}$, $\bar{v}_{i-1}$, $\bar{v}_{i+1}$, $\bar{v}_{i+2}$ in terms of $\bar{v}_i$ and the first four derivatives of $\bar{v}$ at place i. These equations can then be solved to obtain the following formulas, accurate to second order in the difference spacing h:

$$\frac{d\bar{v}}{d\bar{x}} = \frac{\bar{v}_{i+1} - \bar{v}_{i-1}}{2h}$$

$$\frac{d^2\bar{v}}{d\bar{x}^2} = \frac{\bar{v}_{i+1} - 2\bar{v}_i + \bar{v}_{i-1}}{h^2}$$

$$\frac{d^3\bar{v}}{d\bar{x}^2} = \frac{\bar{v}_{i+2} - 2\bar{v}_{i+1} + 2\bar{v}_{i-1} - \dot{\bar{v}}_{i-2}}{2h^3} \tag{B.4}$$

$$\frac{d^4\bar{v}}{d\bar{x}^4} = \frac{\bar{v}_{i+2} - 4\bar{v}_{i+1} + 6\bar{v}_i - 4\bar{v}_{i-1} + \bar{v}_{i-2}}{h^4}$$

With these equations, the boundary conditions of equation (B.3) are easily seen to reduce to the following:

$$\bar{v}_1 = 0, \qquad \bar{v}_2 - 2\bar{v}_1 + \bar{v}_0 = 0$$

$$\bar{v}_5 = 0, \qquad \bar{v}_6 - 2\bar{v}_5 + \bar{v}_4 = 0 \tag{B.5}$$

where fictitious displacements $\bar{v}_0$ and $\bar{v}_6$ outside the beam are introduced. In addition, the differential equation of (B.3) requires at the three interior points that

$$\bar{v}_4 - 4\bar{v}_3 + 6\bar{v}_2 - 4\bar{v}_1 + \bar{v}_0 = -(0.25)^4$$

$$\bar{v}_5 - 4\bar{v}_4 + 6\bar{v}_3 - 4\bar{v}_2 + \bar{v}_1 = -(0.25)^4 \tag{B.6}$$

$$\bar{v}_6 - 4\bar{v}_5 + 6\bar{v}_4 - 4\bar{v}_3 + \bar{v}_2 = -(0.25)^4$$

Solving equations (B.5) and (B.6), we find that the deflections at the interior points are thus determined as

$$\bar{v}_2 = \bar{v}_4 = -0.00976, \qquad \bar{v}_3 = -0.0137 \tag{B.7}$$

These values differ from the actual values by only about 5%.

Computational Molecules. The above one-dimensional finite difference operators may be represented by so-called computational molecules as follows

$$i-2 \qquad i-1 \qquad i \qquad i+1 \qquad i+2$$

$$\left(\frac{d\psi}{dx}\right)_i = \frac{1}{2h} \left\{ \quad \text{(-1)}\!\!-\!\!-\!\!-\!\!\text{(0)}\!\!-\!\!-\!\!-\!\!\text{(1)} \quad \right\}$$

$$\left(\frac{d^2\psi}{dx^2}\right)_i = \frac{1}{h^2} \left\{ \quad \text{(1)}\!\!-\!\!-\!\!\text{(-2)}\!\!-\!\!-\!\!\text{(1)} \quad \right\}$$

$$\left(\frac{d^3\psi}{dx^3}\right)_i = \frac{1}{2h^3} \left\{ \text{(-1)}\!\!-\!\!-\!\!\text{(2)}\!\!-\!\!-\!\!\text{(0)}\!\!-\!\!-\!\!\text{(-2)}\!\!-\!\!-\!\!\text{(1)} \right\}$$

$$\left(\frac{d^4\psi}{dx^4}\right)_i = \frac{1}{h^4} \left\{ \text{(1)}\!\!-\!\!-\!\!\text{(-4)}\!\!-\!\!-\!\!\text{(6)}\!\!-\!\!-\!\!\text{(-4)}\!\!-\!\!-\!\!\text{(1)} \right\}$$

To use these molecules, we simply multiply the enclosed value by the corresponding value of ψ at the indicated location and add or subtract to obtain the appropriate difference expression.

Similarly, in two dimensions we have

$$\left(\frac{\partial\psi}{\partial x}\right)_{i,j} = \frac{1}{2h} \left\{ \text{(-1)}\!\!-\!\!-\!\!\underset{i,j}{\text{(0)}}\!\!-\!\!-\!\!\text{(1)} \right\}$$

$$\vdots$$

$$\left(\frac{\partial^4\psi}{\partial x^4}\right)_{i,j} = \frac{1}{h^4} \left\{ \text{(1)}\!\!-\!\!-\!\!\text{(-4)}\!\!-\!\!-\!\!\underset{i,j}{\text{(6)}}\!\!-\!\!-\!\!\text{(-4)}\!\!-\!\!-\!\!\text{(1)} \right\}$$

and

$$\left(\frac{\partial\psi}{\partial y}\right)_{i,j} = \frac{1}{2h} \left\{ \begin{array}{c} \text{(1)} \\ | \\ \text{(0)}_{i,j} \\ | \\ \text{(-1)} \end{array} \right\}$$

$$\vdots$$

$$\left(\frac{\partial^4\psi}{\partial y^4}\right)_{i,j} = \frac{1}{h^4} \left\{ \begin{array}{c} \text{(1)} \\ | \\ \text{(-4)} \\ | \\ \text{(6)}_{i,j} \\ | \\ \text{(-4)} \\ | \\ \text{(1)} \end{array} \right\}$$

together with, for example,

$$\left(\frac{\partial^2 \psi}{\partial x\,\partial y}\right)_{i,j} = \frac{1}{4h^2}\left\{\begin{array}{c}\text{(stencil)}\end{array}\right\}$$

$$\left(\frac{\partial^4 \psi}{\partial x^2\,\partial y^2}\right)_{i,j} = \frac{1}{h^4}\left\{\begin{array}{c}\text{(stencil)}\end{array}\right\}$$

Two-Dimensional Example. As a simple two-dimensional example, we consider again the strength-of-material formulation of Chapter 5 and examine the deflection of a simply supported square plate of flexural rigidity D and side lengths a subjected to a uniform lateral pressure loading p. Choosing the dimensionless deflection $\bar{w}$ and dimensionless coordinates $\bar{x}$ and $\bar{y}$ as

$$\bar{w} = \frac{wD}{pa^4}, \qquad \bar{x} = \frac{x}{a}, \qquad \bar{y} = \frac{y}{a} \tag{B.8}$$

we obtain the governing equation and boundary conditions as

$$\frac{\partial^4 \bar{w}}{\partial \bar{x}^4} + 2\,\frac{\partial^4 \bar{w}}{\partial \bar{x}^2\,\partial \bar{y}^2} + \frac{\partial^4 \bar{w}}{\partial \bar{y}^4} = -1$$

$$\bar{w} = \frac{\partial^2 \bar{w}}{\partial \bar{x}^2} = 0 \qquad \text{at} \quad \bar{x} = 0,\,1 \tag{B.9}$$

$$\bar{w} = \frac{\partial^2 \bar{w}}{\partial \bar{y}^2} = 0 \qquad \text{at} \quad \bar{y} = 0,\,1$$

Considering now the discrete points shown in Figure B.3 and recognizing that the boundary values $\bar{w}_{0,0}$, $\bar{w}_{1,0}$, etc. are all zero, we may write the governing finite-difference equation for the interior point $(1,1)$ as

$$\bar{w}_{-1,1} + 20\bar{w}_{1,1} + \bar{w}_{3,1} + \bar{w}_{1,3} + \bar{w}_{1,-1} = -(0.5)^4 \tag{B.10}$$

In addition, the remaining boundary conditions provide the equations

$$\begin{aligned}\bar{w}_{-1,1} &= -\bar{w}_{1,1}, & \bar{w}_{3,1} &= -\bar{w}_{1,1} \\ \bar{w}_{1,3} &= -\bar{w}_{1,1}, & \bar{w}_{1,-1} &= -\bar{w}_{1,1}\end{aligned} \tag{B.11}$$

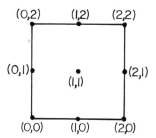

FIGURE B.3.

Hence, on solving these equations, we find the deflection at the center of the plate determined as

$$\bar{w}_{1,1} = -0.00391 \tag{B.12}$$

which is again within about 5% of the actual value.

Of course, with this problem as well as the previous one, if a more refined description of the deflections were required or if strains and stresses were needed (from finite difference estimates of the deflection derivatives), it would be necessary to resort to a much larger network of interior and boundary points.

B.2. Finite Element Method

In applying this method of numerical analysis, the solid body is imagined to be divided into a finite number of simple continuous elements. For each element, the equations of solid mechanics are used to relate the forces and displacements of the element at its contact, or nodal, points. By combining the force–displacement response of each element with equilibrium and boundary conditions of the entire body, the nodal displacements of each element can be determined by solution of algebraic equations and these may then be used to determine strains and stresses within each element.

One-Dimensional Example. We consider the simple problem of determining the stress and displacement in a vertical prismatic bar loaded by its own weight. We divide the body into two elements and label coordinates (or nodal points) as shown in Figure B.4. As an approximation, we regard the distributed weight loading as concentrated forces F_1, F_2, F_3 acting at the coordinates.

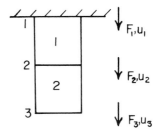

FIGURE B.4.

Considering the first element, we assume a linear displacement distribution of the form

$$u = ax + b \tag{B.13}$$

and evaluate the constants a and b in terms of the end displacements (Figure B.5) to get

$$u = \frac{u_2 - u_1}{h} x + u_1 \tag{B.14}$$

where h denotes the element length. Using the elastic constitutive relation

$$t_{xx} = E e_{xx} = E \frac{\partial u}{\partial x} \tag{B.15}$$

we next find

$$t_{xx} = E \frac{u_2 - u_1}{h} \tag{B.16}$$

so that the force P_2^1 acting on element 1 at coordinate 2 is

$$P_2^1 = \frac{EA}{h} (u_2 - u_1) \tag{B.17}$$

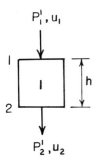

FIGURE B.5.

where A denotes the cross-sectional area of the element. Similarly, we also find the force P_1^1 acting at coordinate 1 of element 1 expressible as

$$P_1^1 = \frac{EA}{h}(u_1 - u_2) \tag{B.18}$$

Equations (B.17) and (B.18) may obviously be written as

$$P_1^1 = k_{11}u_1 + k_{12}u_2, \qquad P_2^1 = k_{21}u_1 + k_{22}u_2 \tag{B.19}$$

where $k_{11} = k_{22} = EA/h$ and $k_{12} = k_{21} = -EA/h$. Corresponding equations for element 2 may likewise be written as

$$P_2^2 = k_{22}u_2 + k_{23}u_3, \qquad P_3^2 = k_{32}u_2 + k_{33}u_3 \tag{B.20}$$

where $k_{22} = k_{33} = EA/h$, $k_{23} = k_{32} = -EA/h$.

If F_1, F_2, F_3 denote externally applied forces (Figure B.3), equilibrium requires that

$$F_1 = P_1^1, \qquad F_2 = P_2^1 + P_2^2, \qquad F_3 = P_3^2 \tag{B.21}$$

Combining equations (B.19)–(B.21), we thus obtain

$$\begin{aligned}
F_1 &= k_{11}u_1 + k_{12}u_2 \\
F_2 &= k_{21}u_1 + 2k_{22}u_2 + k_{23}u_3 \\
F_3 &= k_{32}u_2 + k_{33}u_3
\end{aligned} \tag{B.22}$$

These equations relate the displacement and forces of the overall bar and may be solved for the displacements once applied forces and boundary conditions are specified.

Treating the distributed weight loading as concentrated loads and placing one-half the weight of each element at its ends, we have

$$F_1 = \gamma Ah/2 + F_1^B, \qquad F_2 = \gamma Ah, \qquad F_3 = \gamma Ah/2 \tag{B.23}$$

where γ denotes the weight per unit volume of the bar material and F_1^B denotes the reaction force exerted by the boundary and the fixed end.

Requiring $u_1 = 0$ in accordance with the boundary conditions, we then find from the second and third of equations (B.22) that

$$u_2 = \tfrac{3}{2}\gamma h^2/E, \qquad u_3 = 2\gamma h^2/E \tag{B.24}$$

and from the first, we find

$$F_1 = -\tfrac{3}{2}\gamma hA \tag{B.25}$$

The boundary reaction from equation (B.23) is thus also determined as

$$F_1^B = -2\gamma hA \tag{B.26}$$

It is interesting to note that the above values are identical to those given by exact theory. Having these displacements, we may of course use equation (B.16) and a corresponding one for element 2 to determine the stress in each element. In this case, the stresses will have constant values within each element which are the average values of the actual linearly varying stress distribution.

It is worth noting that equation (B.22) can be written (with the summation convention) as

$$F_i = K_{ij}u_j, \qquad i, j = 1\text{-}3 \tag{B.27}$$

where K_{ij} denotes the *system stiffness matrix* (or array), which may be written in terms of the *element stiffness matrices* k_{ij} as

$$\begin{bmatrix} K_{11} & K_{12} & K_{13} \\ K_{21} & K_{22} & K_{23} \\ K_{31} & K_{32} & K_{33} \end{bmatrix} = \begin{bmatrix} k_{11} & k_{12} & 0 \\ k_{21} & k_{22} & 0 \\ 0 & 0 & 0 \end{bmatrix} + \begin{bmatrix} 0 & 0 & 0 \\ 0 & k_{22} & k_{23} \\ 0 & k_{32} & k_{33} \end{bmatrix} \tag{B.28}$$

$$\text{system} \qquad\qquad \text{element 1} \qquad\qquad \text{element 2}$$

This observation thus shows that the system stiffness matrix can be constructed from the element stiffness matrices by direct addition. Such a procedure, which is easily seen to apply for any linear system of finite elements, is known as the *direct stiffness method*.

Two-Dimensional Example. The finite element method can be applied to two-dimensional problems using a procedure similar to that discussed above. Consider, for example, the thin square plate loaded as shown in Figure B.6. We imagine the plate to be divided into two triangular elements as indicated.

For element 1 (Figure B.7), we assume a linear displacement distribution of the form

$$u = ax + by + c, \qquad v = dx + ey + f \tag{B.29}$$

and evaluate the constants a, b, etc., in terms of the displacements at the coordinates. Recognizing from Figure B.6 the boundary conditions

$$u_1 = v_1 = 0, \qquad u_2 = v_2 = 0 \tag{B.30}$$

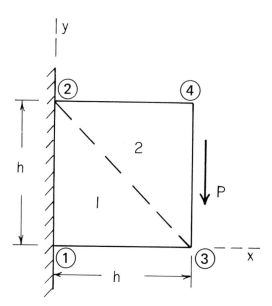

FIGURE. B.6.

we thus find the displacements expressible as

$$u = \frac{u_3}{h} x, \qquad v = \frac{v_3}{h} x \tag{B.31}$$

Assume a state of plane stress in the element; we next determine the stresses as

$$t_{xx} = \frac{E}{1 - v^2} \frac{u_3}{h}$$

$$t_{yy} = \frac{vE}{1 - v^2} \frac{u_3}{h} \tag{B.32}$$

$$t_{xy} = \frac{E}{2(1 + v)} \frac{v_3}{h}$$

where E and v denote Young's modulus and Poisson's ratio, respectively.

Multiplying the stress resultants of Figure B.7 by the area over which they act and placing one-half this force at each end, we may easily determine the forces P^1_{1x}, P^1_{1y}, etc. in terms of the displacements as

$$P^1_{1x} = -k_1 u_3 - k_2 v_3, \qquad P^1_{1y} = -k_3 u_3 - k_2 v_3$$

$$P^1_{2x} = k_2 v_3, \qquad P^1_{2y} = k_3 u_3, \qquad P^1_{3x} = k_1 u_3, \qquad P^1_{3y} = k_2 v_3 \tag{B.33}$$

where

$$k_1 = \frac{Et}{2(1-\nu^2)}, \qquad k_2 = \frac{Et}{4(1+\nu)}, \qquad k_3 = \frac{\nu Et}{2(1-\nu^2)}$$

with t denoting the plate thickness.

In a similar way, we may also determine the following force–deflection response for element 2:

$$P_{2x}^2 = -k_1 u_4 + k_3 v_3 - k_3 v_4$$

$$P_{2y}^2 = k_2 u_3 - k_2 u_4 - k_2 v_4$$

$$P_{3x}^2 = k_2 u_3 - k_2 u_4 - k_2 v_4 \tag{B.34}$$

$$P_{3y}^2 = -k_3 u_4 + k_1 v_3 - k_1 v_4$$

$$P_{4x}^2 = (k_1 + k_2)u_4 - k_2 u_3 - k_3 v_3 + (k_2 + k_3)v_4$$

$$P_{4y}^2 = -k_2 u_3 + (k_2 + k_3)u_4 - k_1 v_3 + (k_1 + k_2)v_4$$

Combining the above equations using the direct stiffness method, we thus obtain a system of equations connecting the applied forces F_{1x}, F_{1y},

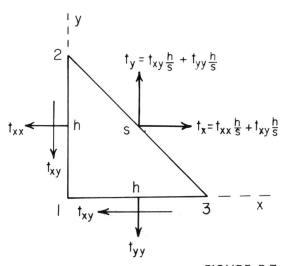

FIGURE B.7.

etc., with the displacements u_1, v_1, etc., in the form

$$F_{1x} = -k_1 u_3 - k_2 v_3$$

$$F_{1y} = -k_3 u_3 - k_2 v_3$$

$$F_{2x} = (k_2 + k_3)v_3 - k_1 u_4 - k_3 v_4$$

$$F_{2y} = (k_2 + k_3)u_3 - k_2 u_4 - k_2 v_4$$

$$F_{3x} = (k_1 + k_2)u_3 - k_2 u_4 - k_2 v_4 \qquad \text{(B.35)}$$

$$F_{3y} = (k_1 + k_2)v_3 - k_3 u_4 - k_1 v_4$$

$$F_{4x} = -k_2 u_3 - k_3 v_3 + (k_1 + k_2)u_4 + (k_2 + k_3)v_4$$

$$F_{4y} = -k_2 u_3 - k_1 v_3 + (k_2 + k_3)u_4 + (k_1 + k_2)v_4$$

On setting

$$F_{3x} = F_{4x} = 0, \qquad F_{3y} = F_{4y} = -P/2 \qquad \text{(B.36)}$$

in accordance with the loading shown in Figure B.6, we may use the last four equations to determine u_3, v_3, u_4, and v_4. The first four equations may then be used to determine the reaction forces at the fixed side and equation (B.32) and its counterpart for element 2 may be used to determine the stress in each element. Taking the case where $v = 0.3$, we find, for example, that the vertical deflections at coordinates 3 and 4 are

$$v_3 = -3.157P/Et, \qquad v_4 = -3.607P/Et \qquad \text{(B.37)}$$

These results underestimate the actual deflection as determined from a plane-stress solution by about 15%. For a more accurate estimate, the plate would, of course, have to be divided into a larger number of elements.

Selected Reading

Crandall, S. H., *Engineering Analysis*. McGraw-Hill Book Co., New York, 1956. Chapter 4 gives an excellent discussion of finite difference methods.

Timoshenko, S. P., and J. N. Goodier, *Theory of Elasticity*. McGraw-Hill Book Co., New York, 1951. The appendix to this book provides a detailed treatment of finite difference methods as applied to problems in elasticity.

Gallagher, R. H., *Finite Element Analysis*. Prentice-Hall, Englewood Cliffs, New Jersey, 1975. A readable treatment of finite element fundamentals.

Desai, C. S., and J. F. Abel, *Introduction to the Finite Element Method*. Van Nostrand Reinhold Co., New York, 1972. Extensive discussion of theory and application of the finite element method.

Index

Abel, J. F., 274
Acceleration:
  components, 23
  definition, 23
  material variables, 23
  spatial variables, 23
Addition:
  tensors, 11
  vectors, 8
Adiabatic deformation, 159
Adiabatic elastic constants, 159
Airy stress function, 109
  polar coordinates, 114
Aluminum, 87, 157
Angle:
  between axes, 5
  between line elements, 25-26
  direction, 4
Angular distortion, 25
Angular momentum, 47
  balance, 50
  local form, 60
Anisotropic, 77
Antisymmetric tensor, 13-14
  vector of, 14
Approximation:
  small deformation, 31
  quasistatic, 165, 166, 197
  uncoupled, 165, 167
Array of tensor, 13
Average stress, 69
Axially symmetric displacement, 116, 177
Axisymmetric bending, 135

Balance laws:
  angular momentum, 50
  energy, 147-149
  linear momentum, 48-50
Bars:
  constants for torsion, 103
  laterally constrained, 211-212
  plastic compression, 234-238
  plastic extension, 231-233
  thermoelastic, 173, 183
  twisting of, 98-103
  vibration of, 183, 213
  viscoelastic, 211, 213
  wave propagation in, 95-98
Beams:
  boundary conditions, 125-126, 143, 180, 207
  concentrated loading, 143
  elastic bending, 110-113, 122-126, 143
  elastic extension, 122-126
  exact solution, 142
  numerical solution, 263-265
  plane stress solution, 110-113
  strength of material treatment, 122-126
  thermal effects, 178-182
  uniform loading, 110-113, 126, 207-209, 263-265
  viscous effects, 205-209
Bell, J. F., 257
Bernoulli-Euler relation, 125
Bernoulli, James, 71
Body force:
  components, 49, 66, 68

Body force (*cont'd*)
  definition, 49
  vector, 49
Boley, B. A., 158, 182
Boundary conditions:
  beams, 123, 125, 143, 180, 207
  bars, 97, 99, 173
  circular hole, 118
  cylinder, 115, 176, 204
  plasticity, 247
  plates, 130, 132-133
  shells, 137, 139
  stress, 63-64
Buckingham pi theorem, 261
Built-in edge, 133, 139
Built-in end, 126, 180, 207
Bulk modulus, 90

Cartesian tensors:
  addition, 11
  definition, 10
  multiplication, 12
  transformation law, 10
Cauchy, Augustin, 1, 71
Cauchy-Green deformation tensor, 28
  inverse of, 46
Cauchy-Green stretch tensor, 28
Cauchy stress law, 57
Cauchy stress tensor, 57
  symmetry of, 61
Cauchy stress vector, 48
Chang, T. S., 45, 68, 89
Characteristic curves, 241
Characteristic directions, 243
Characteristic stress relation, 243
Characteristic velocity relation, 243
Characteristics, 241
Circular hole, 118
Circular rod, 100
Classical elasticity, 71
Clausius-Duhem inequality, 149
Coefficient of thermal expansion, 156
Coleman, B. D., 158
Components:
  acceleration, 23
  body force, 49, 68
  displacement, 22
  normal strain, 35
  polar, 42, 67
  principal strain, 40
  shearing strain, 35

Components (*cont'd*)
  rotation, 28, 33, 36
  strain, 33, 35
  stretch, 28
  stress tensor, 48, 66
  stress vector, 48, 54
  velocity, 22
Compatibility equations, 104-107
Computational molecules, 265-267
Configuration, 73
Conservation of mass, 47
Constitutive relations:
  elastic, 73, 84, 86
  Levy-Mises, 222
  plastic, 217, 221, 222
  Prandtl-Reuss, 221
  strength-of-material, 122, 179, 206
  thermoelastic, 150, 156
  viscoelastic, 187, 191
Contact force, 48
Continuum motion, 21-45
Contraction, 12
Coordinate transformation, 5
Coordinate system:
  Cartesian, 4
  cylindrical polar, 41
Coordinates:
  Eulerian, 21
  initial, 21, 41
  Lagrangian, 21
  material, 21
  particle, 21
  polar, 42
  spatial, 21
Copper, 87, 157
Corotational stress rate, 189, 219
Correspondence principle, 209
Crandall, S. H., 274
Creep, 201
Cushioning mechanics, 262
Cylinder:
  elevated bore temperature, 175
  internal pressure, 115, 227
  Kelvin-Voigt material, 202
  periodic bore pressure, 202
  yielding of, 227
Cylindrical polar coordinates, 41, 67, 113

Damping, 164, 185, 199
Dashpot, 193
Decomposition of deformation, 27-28

Deformation:
  decomposition, 27
  gradients, 23
  homogeneous, 24
  inhomogeneous, 24
  plane, 38
  plastic, 217
  rotation of, 28, 33
  small, 33
  strain of, 33
  stretch of, 25, 28
  tensor components, 26, 27
Deformation tensors, 28
Density, 47
Derivatives:
  material variables, 22
  spatial variable, 23
Desai, C. S., 274
Deviatoric strain, 210
Deviatoric stress, 210
Dilatation, 37
Dimensional analysis, 259
Direct stiffness method, 271
Direction angles, 4
Direction cosines, 4
Displacement:
  axially symmetric, 116
  components, 22, 42
  definition, 22
  gradients, 32
  relative, 33
Distributed loading, 133
Duhamel, J. M. C., 145
Dummy index, 6
Dynamic modulus, 195

Edge:
  simply supported, 132
  built-in, 133, 139
Eigenvalues, 14, 20
Eigenvectors, 14
Elastic concept, 74
Elastic constants:
  adiabatic, 159
  isothermal, 159
  restrictions on, 81
  values of, 87
Elastic constitutive relations, 74, 80, 86
Elastic element, 195
Elastic-plastic solid, 217
Elastic solid, 73

Elastic waves:
  longitudinal, 91, 93
  transverse, 91, 94
Elasticity:
  classical, 71-143
  thermal, 145-184
  viscous, 185-213
Element stiffness matrix, 271
End:
  built-in, 126, 180, 207
  free, 126
  simply supported, 125, 180, 207
Energy, 147
Entropy, 149
Eringen, A. C., 45
Euler, Leonhard, 1, 71
Eulerian finite strain tensor, 46
Eulerian variables, 21
Extension, 39
  beam, 122
  plates, 129
  plastic, 231
  shells, 135

Filaments, 120
Finite difference method, 263-268
Finite element method, 268-274
Finite strain, 45-46
First law of thermodynamics, 147, 149
Flugge, W., 194, 212
Forces:
  body, 49
  stress, 48
Fourier series, 101, 134
Frederick, D., 45, 68, 89
Free end, 126
Frequency of vibration:
  thermal effects, 164
  viscous effects, 199
Fung, Y. C., 68, 141, 194, 212

Gallagher, R. H., 274
General motion, 28-31
General Principles, 1
Geometric interpretation:
  strain, 34
  rotation, 34
Geometry of motion, 21
Glass, 87, 157
Goodier, J. N., 141, 274

Governing equations:
  beams, 126, 180, 207
  elastic, 88, 126, 132, 138
  plastic, 223
  plates, 132
  shells, 138
  thermal, 157, 180
  viscous, 193, 207
Gradients:
  deformation, 23
  displacement, 32
Green's theorem, 52

Hardening, 233
Heat flow, 148
Helmholtz function, 149
Hencky's theorem, 262
Hill, R., 224, 257
Hodge, P. G., 257
Hole in disc, 184
Hole in strained plate, 118
Homogeneous deformation, 24
Homogeneous material, 74
Hooke, Robert, 71
Hypothetical material, 121

Independent dimensionless ratios, 261
Index notation, 4
Infinite rigidity, 120
Inhomogeneous deformation, 24
Inhomogeneous material, 74
Initial coordinates, 21, 41
Initial line element, 25, 28, 34
Inner product, 12
Integrals:
  surface and volume, 52
  time derivative of, 50
Internal pressure, 115, 227
Invariants, 219, 221
Inverse of deformation gradients, 46
Inviscid response, 193
Isotropic material, 77

Jeffreys, H., 19
Johns, D. G., 182

Kelvin, Lord, 185
Kelvin-Voigt solid, 190
Kinematics, 21-45
Kinetic energy, 148
Koiter, W. T., 141

Kolsky, H., 141
Kronecker delta, 7

Lagrangian strain tensor, 45
Lagrangian variables, 21
Lamé constants, 86
Langhaar, H. L., 262
Laterally constrained bar, 211
Lee, E. H., 257
Left Cauchy-Green tensors, 28
Leigh, D. C., 89
Levy, M., 215
Levy-Mises relations, 222
Line element, 26, 30
  angular distortion, 25
  stretch, 25
Linear elastic deformation, 80, 88
Linear momentum:
  balance, 48
  local form, 59
Linear thermoelastic deformation, 153, 157
Linear viscoelastic deformation, 189, 193
Local form:
  mass conservation, 58
  linear momentum, 59
  angular momentum, 60
Lockett, F. J., 194
Long, R. R., 19, 68, 89, 141
Longitudinal wave, 91
Longitudinal vibration, 161
Love, A. E. H., 89

Malvern, L. E., 19, 45, 68, 89, 124, 158
Mass conservation, 47-48
  local form, 58
Mass density, 47
Material constants, 261
Material description, 22
Material variables, 21
Materials:
  homogeneous, 74
  indifference principle, 74, 152, 188, 218
  inhomogeneous, 74
  isotropic, 84
  motion, 28
  stability condition, 213
  symmetry restrictions, 77
  symmetries, 82
Matrix, 211
Maxwell, James, 215
Maxwell-Mises yield condition, 220

Maxwell substance, 190
Moment, 50
Moment-curvature relation, 125
Momentum balance:
　angular, 67
　linear, 48
　local forms, 60, 61
　small deformation, 66
Motion of material, 28
Multiplication:
　tensors, 12
　vectors, 8

Natural configuration, 73
Navier, Louis, 71
Navier's equations, 88
Negligible elastic deformation, 222
Neumann, Franz, 145
Noll, W., 158
Normal strain components, 35
Notation:
　index, 4
　unabridged, 40, 66
Numerical methods, 263
Numerical solutions, 247, 263-274

Orthogonality conditions, 26
Orthogonality relations, 7
Orthotropic material, 83
Outer product, 12

Parkus, H., 182
Particle, 21
Periodic variation, 164
Pi theorem, 261
Plane shock, 159
Plane deformation, 38
Plane strain, 107, 167, 234, 238
Plane stress, 61, 70, 107, 167
Plastic deformation, 217
Plasticity theory, 217
Plastics, 187
Plate, 118, 129, 183
　numerical solution, 267
Poisson's ratio, 87
Polar coordinates:
　displacement components, 42
　plane strain, 113, 174
　plane stress, 113, 174
　rotation components, 42, 44
　strain components, 42, 44
　stress components, 68

Polar decomposition, 28
Polymers, 187
Prager, W., 257
Prandtl, L., 215
Prandtl-Reuss relations, 221
Prandtl's cycloid solution, 258
Prandtl's formulation, 142
Pressure, 61
Principal axes, 16, 40, 62
Principal strains, 40
Principal stresses, 63
Principle of material indifference, 74, 152,
　188, 218
Principle of superposition, 90, 143
Products, 8, 12
Pure dilatation, 37
Pure shear, 39

Quasilinear response, 220
Quasistatic, 197
Quotient rule, 12

Range convention, 5
Rankin-Hugoniot relations, 159
Rate of rotation, 188, 195
Rate of strain, 188, 195
Rayleigh, Lord, 185
Recoverable deformation, 217
Rectangular bar, 100
Rectangular plates, 129
Reference configuration, 73
Reflection of waves, 96
Relative displacement, 33
Reuss, E., 215
Right Cauchy-Green tensors, 28
Rigid rotation, 29, 30
Rigid-body motion, 26
Rigid-plastic solid, 222, 238-257
Rod, 86, 98, 228
Rotation, 21
　components, 36
　displacement relations, 44
　geometric interpretation, 34
　small deformation, 33
　tensor, 28
　vector, 33
Rough wedge penetration, 258

Saint-Venant, Barre de, 71, 215
Saint-Venant's principle, 103
Saint-Venant's warping function, 98

Scalar, 3
Scalar product, 8
Scale modeling, 259
Scaling law, 260
Second law of thermodynamics, 149
Shear, 39
Shear lines, 244
Shearing strain, 35
Shells, 135
Shock wave, 69, 158-159
Similitude, 259
Simmonds, J. G., 141
Simple extension, 39
Simple fan, 253
Simply supported edge, 132
Simply supported end, 125, 180, 207
Slip line, 244
Slip line theory, 243, 247
Small deformations:
    approximation, 31, 65
    elastic relations, 80
    linear momentum, 67
    mass conservation, 66, 68
    rotation tensor, 34
    strain tensor, 34
    thermoelastic relations, 153
    viscoelastic relations, 189
Small temperature changes, 153
Soper, W. G., 262
Soper's scaling law, 262
Spatial coordinates, 21
Spatial description, 22
Specific heat, 156
Standard viscoelastic solid, 187, 190
Statically determinate, 264
Stationary shock, 69, 159
Steels, 87, 157
Strain:
    components, 33, 41
    energy function, 81
    examples, 37
    finite, 45-46
    geometric interpretation, 34
    hardening, 233
    plane, 107, 113
    polar coordinates, 42
    principal axes, 40
    relaxation, 202
    tensor, 33
    transformation, 39
Strength of materials, 120, 178, 205

Stress:
    boundary conditions, 63
    components, 57, 60
    characteristic curve, 243
    characteristic relation, 243
    examples, 61
    mean, 220
    plane, 61, 107, 113
    polar coordinates, 68
    rate, 187, 189
    relaxation, 202
    tensor, 56
    vector, 54
    wave, 94, 96
    yield condition, 217, 220
Stretch, 29, 30
    components, 45
    tensors, 28
Summation convention, 6
Superposition, 90, 143
Symmetric tensor, 13, 16
System stiffness matrix, 271

Temperature, 149
Temperature changes, 153
Tensors:
    antisymmetric, 14
    array, 13
    definition, 10
    symmetric, 13, 16
    transformation law, 10-11
Theory of characteristics, 240
Thermal bending, 180
Thermal conductivity, 156
Thermal constants, 156
    values of, 157
Thermal stresses, 171, 178, 182
Thermodynamics:
    first law, 147
    second law, 149
Thermoelastic damping, 164
Thermoelastic solid, 150
Thermoelastic vibrations, 161
Thick-walled cylinder, 115
Thin beam, 110
Thin plates, 129
Thin shells, 135
Timoshenko, S. P., 141, 182, 274
Torque, 50
Torsion constants, 103
Toupin, R. A., 45, 69

Transformation of components, 5, 11
Translation, 29, 30
Transmission of waves, 96
Transverse wave, 94
Transverse vibration, 197
Tresca, H., 215, 221
Tresca's stress condition, 257
Truesdell, C., 45, 69, 89
Tube, 225
Tupper, S. J., 257
Twisting of rods, 98

Unabridged notation, 66
Uncoupled thermoelasticity, 165
Uniaxial constitutive relation, 191
Uniaxial response, 199
Uniform loading, 110, 126, 207
Uniform pressure loading, 139
Unit eigenvector, 15
Unit extension, 35, 46
Unit normal, 48
Unit vectors, 8

Values of elastic constants, 87
Values of thermal constants, 157
Variables:
  Eulerian, 21
  Lagrangian, 21
  material, 21
  spatial, 21
Vector, 3
  addition, 8
  components, 4
  definition, 3
  magnitude of, 4
  multiplication, 8, 9
  product, 9
  rotation, 33
  transformation law, 5
  unit magnitude, 8
Velocity:
  characteristic curve, 243
  characteristic relation, 243

Velocity (*cont'd*)
  components, 22
  definition, 22
  material description, 23
  spatial description, 23
Vibration:
  thermoelastic solid, 161
  viscoelastic solid, 197
Viscoelastic correspondence principle, 209
Viscoelastic solid, 187
  uniaxial response, 199
  vibration, 197
Viscous elasticity, 185
Viscous element, 195
Viscous lag angle, 195
Voigt, Woldemar, 185
Volume element motion, 29-31
Volume integral:
  time derivative, 50
  transformation, 52
Von Mises, R., 215

Warping function, 98, 141
Wave:
  elastic, 91
  longitudinal, 93
  propagation, 95
  reflection, 96
  stress, 94
  transmission, 96
  transverse, 94
Wedge penetration, 254
Weiner, J. H., 158, 182
Woinowsky-Krieger, S., 141
Work, 89, 147

Yield condition, 217, 218, 220
Yield stress, 220
Young's modulus, 87

Zero body forces, 92
Zero mean stress, 220